Mapping FISH HABITATS

Teacher's Guide

Grades 6–10

Skills
Observing, Classifying, Mapping, Analyzing Data, Experimenting, Making Inferences

Concepts
Habitat Requirements, Home Range, Animal Territories, Research Techniques of Biologists, Monitoring an Aquarium Ecosystem, Humane Treatment of Animals

Themes
Systems & Interactions, Stability, Patterns of Change, Evolution, Structure, Diversity & Unity

WHAT ARE THEMES?
Themes can be seen as major, recurring ideas that provide a framework for the science curriculum. For more on what GEMS means by themes, please see page... 47

Nature of Science and Mathematics
Scientific Community, Interdisciplinary, Cooperative Efforts, Creativity & Constraints, Theory-Based and Testable, Changing Nature of Facts and Theories, Objectivity & Ethics, Science & Society, Real-Life Applications

Time
Three 30- to 45-minute sessions plus daily observations and follow-up sessions

Katharine Barrett
Cary I. Sneider

LHS GEMS
Great Explorations in Math and Science (GEMS)
Lawrence Hall of Science
University of California at Berkeley

Illustrations
Carol Bevilacqua

Photographs
Richard Hoyt
Bob Neumann

Lawrence Hall of Science, University of California, Berkeley, CA 94720. Chairman: Glenn T. Seaborg; Director: Marian C. Diamond

Publication was made possible by grants from the A.W. Mellon Foundation and the Carnegie Corporation of New York. This support does not imply responsibility for statements or views expressed in publications of the GEMS program. GEMS also gratefully acknowledges the contribution of word processing equipment from Apple Computer, Inc. Under a grant from the National Science Foundation, GEMS Leader's Workshops have been held across the country. For further information on GEMS leadership opportunities, please contact GEMS at the address and phone number below.

International Standard Book Number: 0-912511-61-3

COMMENTS WELCOME

Great Explorations in Math and Science (GEMS) is an ongoing curriculum development project. GEMS guides are revised periodically, to incorporate teacher comments and new approaches. We welcome your criticisms, suggestions, helpful hints, and any anecdotes about your experience presenting GEMS activities. Your suggestions will be reviewed each time a GEMS guide is revised. Please send your comments to: GEMS Revisions, c/o Lawrence Hall of Science, University of California, Berkeley, CA 94720. The phone number is (510) 642-7771.

Great Explorations in Math and Science (GEMS) Program

The Lawrence Hall of Science (LHS) is a public science center on the University of California at Berkeley campus. LHS offers a full program of activities for the public, including workshops and classes, exhibits, films, lectures, and special events. LHS is also a center for teacher education and curriculum research and development.

Over the years, LHS staff have developed a multitude of activities, assembly programs, classes, and interactive exhibits. These programs have proven to be successful at the Hall and should be useful to schools, other science centers, museums, and community groups. A number of these guided-discovery activities have been published under the Great Explorations in Math and Science (GEMS) title, after an extensive refinement process that includes classroom testing of trial versions, modifications to ensure the use of easy-to-obtain materials, and carefully written and edited step-by-step instructions and background information to allow presentation by teachers without special background in mathematics or science.

LHS GEMS

Staff
Glenn T. Seaborg, Principal Investigator
Jacqueline Barber, Director
Cary Sneider, Curriculum Specialist
Katharine Barrett, John Erickson, Jaine Kopp, Kimi Hosoume, Laura Lowell, Linda Lipner, Carolyn Willard, Staff Development Specialists
Jan M. Goodman, Mathematics Consultant
Cynthia Ashley, Administrative Coordinator
Gabriela Solomon, Distribution Coordinator
Lisa Haderlie Baker, Art Director
Carol Bevilacqua and Lisa Klofkorn, Designers
Lincoln Bergman and Kay Fairwell, Editors

Contributing Authors
Leigh Agler
Jeremy Ahouse
Jacqueline Barber
Katharine Barrett
Lincoln Bergman
Marion E. Buegler
David Buller
Linda De Lucchi
Jean Echols
Alan Gould
Cheryll Hawthorne
Sue Jagoda
Jefferey Kaufmann
Robert C. Knott
Larry Malone
Cary I. Sneider
Elizabeth Stage
Jennifer Meux White

Reviewers

We would like to thank the following educators who reviewed, tested, or coordinated the reviewing of this series of GEMS materials in manuscript form. Their critical comments and recommendations contributed significantly to these GEMS publications. Their participation does not necessarily imply endorsement of the GEMS program.

ARIZONA

David P. Anderson
Royal Palm Junior High School, Phoenix

Joanne Anger
John Jacobs Elementary School, Phoenix

Cheri Balkenbush
Shaw Butte Elementary School, Phoenix

Flo-Ann Barwick Campbell
Mountain Sky Junior High School, Phoenix

Sandra Caldwell
Lakeview Elementary School, Phoenix

Richard Clark*
Washington School District, Phoenix

Kathy Culbertson
Moon Mountain Elementary School, Phoenix

Don Diller
Sunnyslope Elementary School, Phoenix

Barbara G. Elliot
Tumbleweed Elementary School, Phoenix

Joseph M. Farrier
Desert Foothills Junior High School, Phoenix

Mary Anne French
Moon Mountain Elementary School, Phoenix

Leo H. Hamlet
Desert View Elementary School, Phoenix

Elaine Hardt
Sunnyslope Elementary School, Phoenix

Walter Carroll Hart
Desert View Elementary School, Phoenix

Tim Huff
Sunnyslope Elementary School, Phoenix

Stephen H. Kleinz
Desert Foothills Junior High School, Phoenix

Alison Lamborghini
Orangewood Elementary School, Phoenix

Karen Lee
Moon Mountain Elementary School, Phoenix

George Lewis
Sweetwater Elementary School, Phoenix

Tom Lutz
Palo Verde Junior High School, Phoenix

Midori Mits
Sunset Elementary School, Phoenix

Brenda Pierce
Cholla Junior High School, Phoenix

Sue Poe
Palo Verde Junior High School, Phoenix

Robert C. Rose
Sweetwater Elementary School, Phoenix

Liz Sandberg
Desert Foothills Junior High School, Phoenix

Jacque Sniffen
Chaparral Elementary School, Phoenix

Rebecca Staley
John Jacobs Elementary School, Phoenix

Sandra Stanley
Manzanita Elementary School, Phoenix

Chris Starr
Sunset Elementary School, Phoenix

Karen R. Stock
Tumbleweed Elementary School, Phoenix

Charri L. Strong
Mountain Sky Junior High School, Phoenix

Shirley Vojtko
Cholla Junior High School, Phoenix

K. Dollar Wroughton
John Jacobs Elementary School, Phoenix

CALIFORNIA

Carolyn R. Adams
Washington Primary School, Berkeley

Judith Adler*
Walnut Heights Elementary School, Walnut Creek

Gretchen P. Anderson
Buena Vista Elementary School, Walnut Creek

Beverly Braxton
Columbus Intermediate School, Berkeley

Dorothy Brown
Cave Elementary School, Vallejo

Christa Buckingham
Seven Hills Intermediate School, Walnut Creek

Elizabeth Burch
Sleepy Hollow Elementary School, Orinda

Katharine V. Chapple
Walnut Heights Elementary School, Walnut Creek

Linda Clar
Walnut Heights Elementary School, Walnut Creek

Gail E. Clarke
The Dorris-Eaton School, Walnut Creek

Sara J. Danielson
Albany Middle School, Albany

Robin Davis
Albany Middle School, Albany

Margaret Dreyfus
Walnut Heights Elementary School, Walnut Creek

Jose Franco
Columbus Intermediate School, Berkeley

Elaine Gallaher
Sleepy Hollow Elementary School, Orinda

Ann Gilbert
Columbus Intermediate School, Berkeley

Gretchen Gillfillan
Sleepy Hollow Elementary School, Orinda

Brenda S.K. Goo
Cave Elementary School, Vallejo

Beverly Kroske Grunder
Indian Valley Elementary School, Walnut Creek

Kenneth M. Guthrie
Walnut Creek Intermediate School, Walnut Creek

Joan Hedges
Walnut Heights Elementary School, Walnut Creek

Corrine Howard
Washington Elementary School, Berkeley

Janet Kay Howard
Sleepy Hollow Elementary School, Orinda

Gail Isserman
Murwood Elementary School, Walnut Creek

Carol Jensen
Columbus Intermediate School, Berkeley

Dave Johnson
Cave Elementary School, Vallejo

Kathy Jones
Cave Elementary School, Vallejo

Dayle Kerstad*
Cave Elementary School, Vallejo

Diane Knickerbocker
Indian Valley Elementary School, Walnut Creek

Joan P. Kunz
Walnut Heights Elementary School, Walnut Creek

Randy Lam
Los Cerros Intermediate School, Danville

Philip R. Loggins
Sleepy Hollow Elementary School, Orinda

Jack McFarland
Albany Middle School, Albany

Betty Maddox
Walnut Heights Elementary School, Walnut Creek

Chiyomi Masuda
Columbus Intermediate School, Berkeley

Katy Miles
Albany Middle School, Albany

Lin Morehouse*
Sleepy Hollow Elementary Schoool, Orinda

Marv Moss
Sleepy Hollow Elementary School, Orinda

Tina L. Neivelt
Cave Elementary School, Vallejo

Neil Nelson
Cave Elementary School, Vallejo

Joyce Noakes
Valle Verde Elementary School, Walnut Creek

Jill Norris
Sleepy Hollow Elementary School, Orinda

Janet Obata
Albany Middle School, Albany

Patrick Pase
Los Cerros Intermediate School, Danville

Geraldine Piglowski
Cave Elementary School, Vallejo

Susan Power
Albany Middle School, Albany

Louise Rasmussen
Albany Middle School, Albany

Jan Rayder
Columbus Intermediate School, Berkeley

Masha Rosenthal
Sleepy Hollow Elementary School, Orinda

Carol Rutherford
Cave Elementary School, Vallejo

Jim Salak
Cave Elementary School, Vallejo

Constance M. Schulte
Seven Hills Intermediate School, Walnut Creek

Robert Shogren*
Albany Middle School, Albany

Kay L. Sorg*
Albany Middle School, Albany

Marc Tatar
University of California Gifted Program, Berkeley

Mary E. Welte
Sleepy Hollow Elementary School, Orinda

Carol Whitmore-Waldron
Cave Elementary School, Vallejo

Vernola J. Williams
Albany Middle School, Albany

Carolyn Willard*
Columbus Intermediate School, Berkeley

Mary Yonekawa
The Dorris-Eaton School, Walnut Creek

KENTUCKY

Joyce M. Anderson
Carrithers Middle School, Louisville

Susan H. Baker
Museum of History and Science, Louisville

Carol Earle Black
Highland Middle School, Louisville

April B. Bond
Rangeland Elementary School, Louisville

Sue M. Brown
Newburg Middle School, Louisville

Donna Ross Butler
Carrithers Middle School, Louisville

Stacey Cade
Sacred Heart Model School, Louisville

Sister Catherine, O.S.U.
Sacred Heart Model School, Louisville

Judith Kelley Dolt
Gavin H. Cochran Elementary School, Louisville

Elizabeth Dudley
Carrithers Middle School, Louisville

Jeanne Flowers
Sacred Heart Model School, Louisville

Karen Fowler
Carrithers Middle School, Louisville

Laura Hansen
Sacred Heart Model School, Louisville

Sandy Hill-Binkley
Museum of History and Science, Louisville

Deborah M. Hornback
Museum of History and Science, Louisville

Patricia A. Hulak
Newburg Middle School, Louisville

Rose Isetti
Museum of History and Science, Louisville

Mary Ann M. Kent
Sacred Heart Model School, Louisville

James D. Kramer
Gavin H. Cochran Elementary School, Louisville

Sheneda Little
Gavin H. Cochran Elementary School, Louisville

Brenda W. Logan
Newburg Middle School, Louisville

Amy S. Lowen*
Museum of History and Science, Louisville

Mary Louise Marshall
Breckinridge Elementary School, Louisville

Theresa H. Mattei*
Museum of History and Science, Louisville

Judy Reibel
Highland Middle School, Louisville

Pamela R. Record
Highland Middle School, Louisville

Margie Reed
Carrithers Middle School, Louisville

Donna Rice
Carrithers Middle School, Louisville

Ken Rosenbaum
Jefferson County Public Schools, Louisville

Edna Schoenbaechler
Museum of History and Science, Louisville

Karen Schoenbaechler
Museum of History and Science, Louisville

Deborah G. Semenick
Breckinridge Elementary School, Louisville

Dr. William McLean Sudduth*
Museum of History and Science, Louisville

Rhonda H. Swart
Carrithers Middle School, Louisville

Arlene S. Tabor
Gavin H. Cochran Elementary School, Louisville

Carla M. Taylor
Museum of History and Science, Louisville

Carol A. Trussell
Rangeland Elementary School, Louisville

Janet W. Varon
Newburg Middle School, Louisville

MICHIGAN

Glen Blinn
Harper Creek High School, Battle Creek

Douglas M. Bollone
Kelloggsville Junior High School, Wyoming

Sharon Christensen*
Delton-Kellogg Middle School, Delton

Ruther M. Conner
Parchment Middle School, Kalamazoo

Stirling Fenner
Gull Lake Middle School, Hickory Corners

Dr. Alonzo Hannaford*
Western Michigan University, Kalamazoo

Barbara Hannaford
The Gagie School, Kalamazoo

Duane Hornbeck
St. Joseph Elementary School, Kalamazoo

Mary M. Howard
The Gagie School, Kalamazoo

Diane Hartman Larsen
Plainwell Middle School, Plainwell

Miriam Hughes
Parchment Middle School, Kalamazoo

Dr. Phillip T. Larsen*
Western Michigan University, Kalamazoo

David M. McDill
Harper Creek High School, Battle Creek

Sue J. Molter
Dowagiac Union High School, Dowagiac

Julie Northrop
South Junior High School, Kalamazoo

Judith O'Brien
Dowagiac Union High School, Dowagiac

Rebecca Penney
Harper Creek High School, Battle Creek

Susan C. Popp
Riverside Elementary School, Constantine

Brenda Potts
Riverside Elementary School, Constantine

Karen Prater
St. Joseph Elementary School, Kalamazoo

Joel Schuitema
Woodland Elementary School, Portage

Pete Vunovich
Harper Creek Junior High School, Battle Creek

Beverly E. Wrubel
Woodland Elementary School, Portage

NEW YORK

Frances P. Bargamian
Trinity Elementary School, New Rochelle

Barbara Carter
Jefferson Elementary School, New Rochelle

Ann C. Faude
Heathcote Elementary School, Scarsdale

Steven T. Frantz
Heathcote Elementary School, Scarsdale

Alice A. Gaskin
Edgewood Elementary School, Scarsdale

Harriet Glick
Ward Elementary School, New Rochelle

Richard Golden*
Barnard School, New Rochelle

Seymour Golden
Albert Leonard Junior High School, New Rochelle

Don Grant
Isaac E. Young Junior High School, New Rochelle

Marybeth Greco
Heathcote Elementary School, Scarsdale

Peter C. Haupt
Fox Meadow Elementary School, Scarsdale

Tema Kaufman
Edgewood Elementary School, Scarsdale

Donna MacCrae
Webster Magnet Elementary School, New Rochelle

Dorothy T. McElroy
Edgewood Elementary School, Scarsdale

Mary Jane Motl
Greenacres Elementary School, Scarsdale

Tom Mullen
Jefferson Elementary School, New Rochelle

Robert Nebens
Ward Elementary School, New Rochelle

Eileen L. Paolicelli
Ward Elementary School, New Rochelle

Donna Pentaleri
Heathcote Elementary School, Scarsdale

Dr. John V. Pozzi*
City School District of New Rochelle, New Rochelle

John J. Russo
Ward Elementary School, New Rochelle

Bruce H. Seiden
Webster Magnet Elementary School, New Rochelle

David B. Selleck
Albert Leonard Junior High School, New Rochelle

Lovelle Stancarone
Trinity Elementary School, New Rochelle

Tina Sudak
Ward Elementary School, New Rochelle

Julia Taibi
Davis Elementary School, New Rochelle

Kathy Vajda
Webster Magnet Elementary School, New Rochelle

Charles B. Yochim
Davis Elementary School, New Rochelle

Bruce D. Zeller
Isaac E. Young Junior High School, New Rochelle

DENMARK

Dr. Erik W. Thulstrup
Royal Danish School of Educational Studies, Copenhagen

***Trial test coordinators**

Contents

Acknowledgments

This activity was developed at the Lawrence Hall of Science by Jefferey Kaufmann, as part of a series of science courses on animal behavior. Gigi Bridges and Mary Glanville contributed ideas for improving the activities during pilot testing. Bea Boffen and Susan Brady have been especially helpful in providing advice for setting up and maintaining an aquarium.

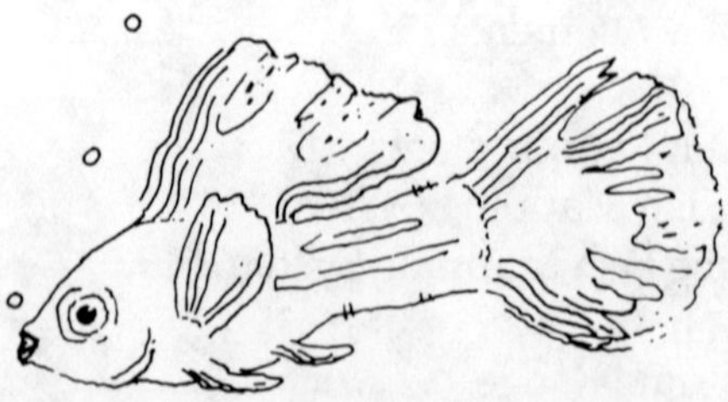

Introduction

We usually identify different kinds of fish by their appearance. In these activities, your students discover that fish also have behavioral differences. They may find, for example, that different species of fish tend to swim in different parts of the aquarium, or react differently to changes in the positions of rocks and plants. Your students may even observe fish defending their territories, or notice that some individuals are more dominant than others. These different behaviors provide clues about what the various fish need to survive and how they interact with each other and their environment.

While many children are fascinated by the graceful movements of fish, they are unlikely to learn much about fish behavior through casual observations alone. The activities in this unit will help you guide your students in **observing** and **recording** fish movements systematically, **experimenting** by changing one element of the environment to see if it causes changes in fish behavior, and then **analyzing** and **interpreting** the data to draw conclusions. Since the activities are based on methods actually used by aquatic biologists, your students will learn about the work professional scientists do.

By using an aquarium as a research laboratory, you will be able to make important concepts "come alive" in ways that textbooks cannot. Your students will learn that an ***ecosystem*** includes all of the plants and animals in an area, along with *all elements* of their non-living environment, such as soil, water, air, and sunlight. They will observe the ***home range***, or area used by each fish, within the ecosystem of the aquarium. They will learn that the ***habitat*** of a given animal includes the area where it lives, as well as the things and conditions that it needs to survive. Your students may also see examples of fish defending their ***territory,*** that part of the habitat that some fish protect from other animals. Depending on the concepts covered by your science textbook, you may also wish to use the aquarium to illustrate the oxygen-carbon dioxide cycle and the nitrogen cycle.

If you already have an aquarium in your classroom, you have completed most of the work needed to prepare for this unit. If not, a brief background section on "Setting up an Aquarium" gives basic information, and a "Resources" section suggests a number of good books on the topic.

As with all laboratory experiments, only some of the conditions of the natural environment can be replicated in your classroom aquarium. Make sure your students understand that the animals are not necessarily behaving as they would in their natural environment.

Summary Outlines are provided to help you guide your students through these activities in an organized way. Student worksheets appear immediately following the session in which they are needed. Removable copies of the student worksheets can be found in the back of the booklet.

While the activities in this guide may be used as an introduction to the study of animal behavior, or the concept of habitat, a more extensive unit might include these three GEMS Teacher's Guides, presented in the following order:

Animals In Action provides a good introduction to animal behavior, and allows you to emphasize the difference between observations and inferences.

Mapping Animal Movements introduces systematic observation and sampling methods, as applied to insects and mammals.

Mapping Fish Habitats extends systematic sampling methods to fish in classroom aquariums and encourages students to design their own experiments.

Activity 1: Setting up the Fish Tank

Overview

In this activity, your students design the environment for their fish mapping activities by placing rocks, plants, and fish in the aquarium. The purpose of the activity is to stimulate the students' thinking about how the rocks and plants will affect fish behavior, and to recognize that the aquarium is an ***aquatic ecosystem.***

The students also measure environmental characteristics of the aquarium (temperature, pH, water level, and algae growth), and find out how to adjust those characteristics in order to maintain healthy fish. This part of the activity is very important, as it helps students accept responsibility for the welfare of animals that they use in experiments.

Time Frame

Introducing the Project	5 minutes
Making and Monitoring the Ecosystem	20 minutes
Adding and Observing the Fish	20 minutes

What You Need

For the class:

- ☐ 1 aquarium, set up for tropical fish (See "Setting up an Aquarium" on page 39, for a detailed list of materials and background information.)
- ☐ 5 or 6 fish of three different species (See "Getting Ready" on page 6 for suggestions)
- ☐ 2 sheets of white paper (12″ × 18″ or larger)
- ☐ 1 pH kit for aquariums
- ☐ 1 thermometer
- ☐ 1 ruler
- ☐ 1 roll of masking tape
- ☐ 1 container of flake fish food (See page 44 for more information on feeding and possible use of pellet, tablet, and live food.)

For each team of 4–6 students:

- ☐ "Fish Facts" information sheet (See "Getting Ready" on page 6.)

For each student:

- ☐ 1 piece of lined paper
- ☐ 1 pencil

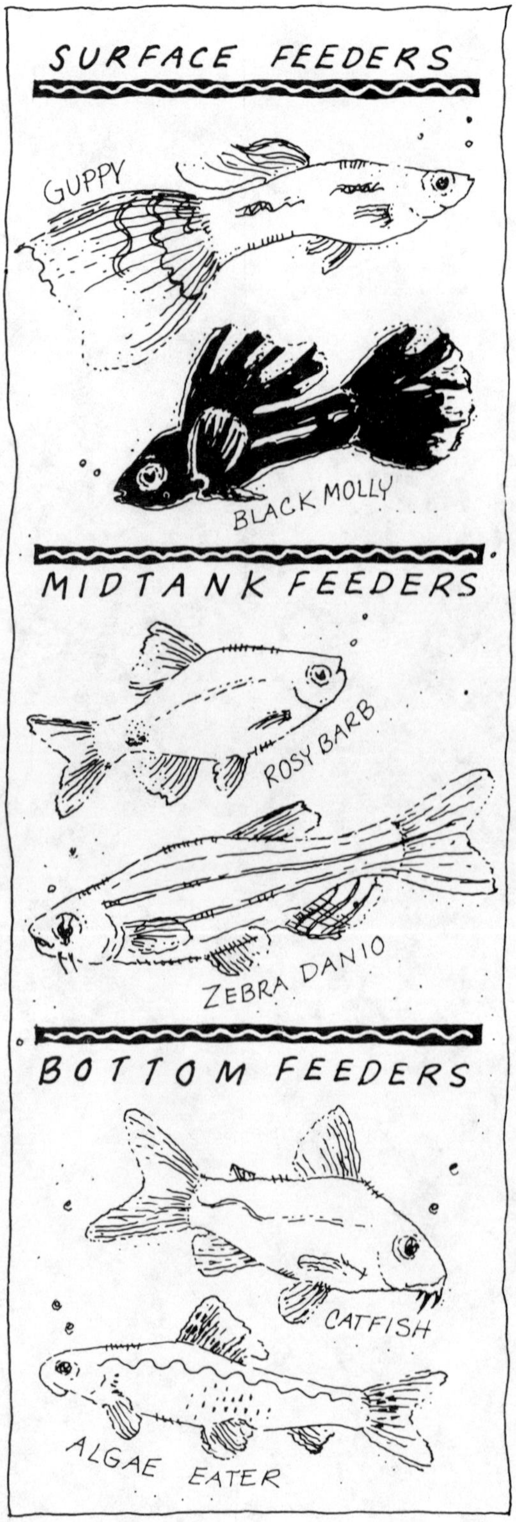

Getting Ready

1. If you are setting up an aquarium for the first time, refer to "Setting up an Aquarium" on page 39. Set it up at least one week before starting the activity to allow conditions in the aquarium to stabilize before the fish are added.

2. For these activities, you will need **at least three species of fish, and a total of five or six individuals.** One of the species should be a surface feeder, such as guppies or black mollies. Select a second species that spends time near the center of the aquarium, such as rosy barbs or angelfish. A third species should be a bottom feeder, such as catfish or algae eaters. By selecting males and females, or individuals of different sizes, you will increase the opportunities for observing interactions related to breeding, territory, and dominance. If you already have an aquarium with more than six fish, there is no need to remove any of them, but plan to limit the number that students will map to five or six, as that is the largest number of fish that a team will be able to keep track of or monitor.

3. Prepare a "Fish Facts" information sheet. On a single sheet of paper, write the name of each kind of fish that you have selected for these activities, and draw a simple sketch so the students can learn to tell the fish apart. Add a few sentences of information about each kind of fish. (If you select guppies, rosy barbs, and catfish, you may copy the sample provided on page 12. Information on other fish is available from most aquarium guides, or from pet store operators.) Make one copy of the "Fish Facts" information sheet for each team.

4. Make an Aquarium Chart for recording measurements of environmental conditions within the fish tank on a weekly basis. The chart should include: temperature, pH, water level, and algae growth.

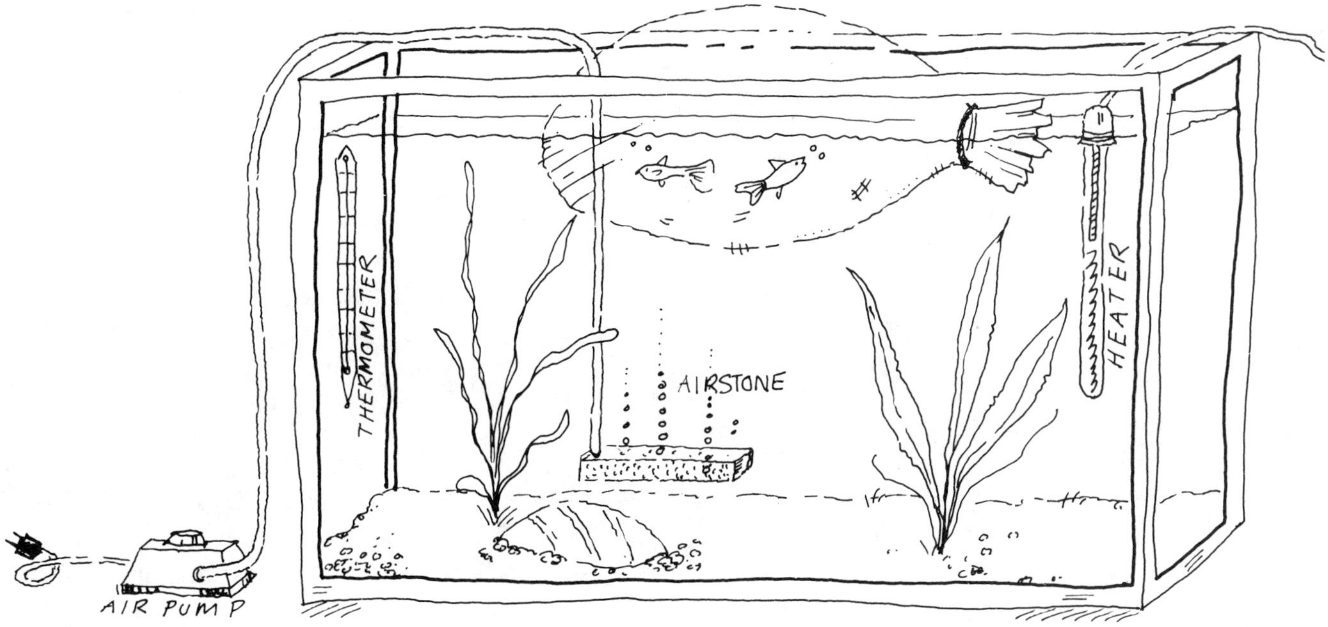

5. Make a Daily Feeding Schedule to designate the days that different teams are responsible for feeding the fish. Post the Aquarium Chart and the Daily Feeding Schedule on the wall near the aquarium.

6. Decide on a few rocks and plants that the students can move or add to the aquarium during the introductory activity, "Making and Monitoring the Ecosystem."

7. If you are adding new fish, arrange to purchase them close to the day when your students can put them into the aquarium. Fish are usually transported from the pet shop in a plastic bag with lots of water and air. Float the bags in a holding tank or the aquarium until the temperatures equalize (10–20 minutes).

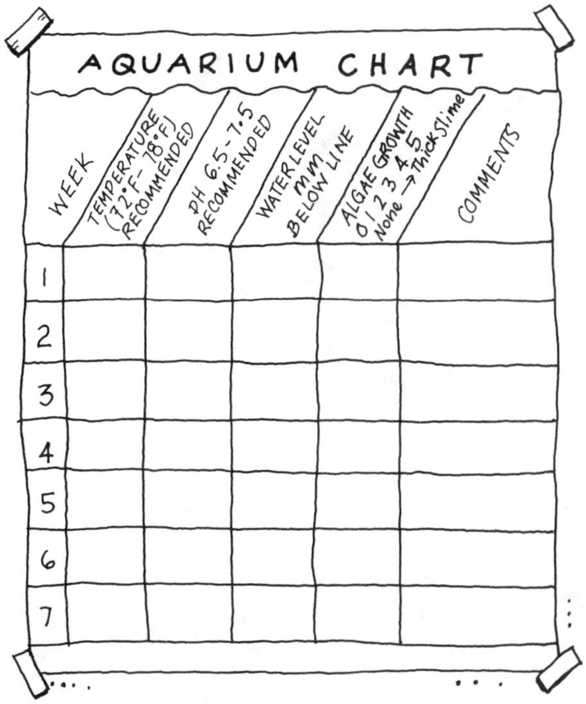

AQUARIUM CHART FOR RECORDING WEEKLY MEASUREMENTS OF TEMPERATURE, pH, WATER LEVEL & ALGAE GROWTH.

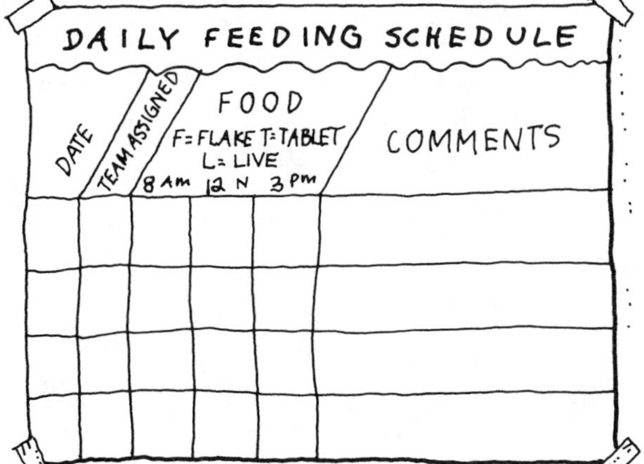

Introducing the Project

1. Tell the students that in the next few days and weeks they will form teams and conduct experiments using the aquarium as an indoor research laboratory to find out how different kinds of fish interact, and in which areas of the aquarium they tend to spend the most time.

2. Point out that the aquarium is a small ***ecosystem.*** *Eco* comes from the Greek word *oikos,* meaning house. Define ***system*** as a group of interacting parts. An ecosystem includes all of the plants and animals in an area, as well as other environmental conditions, such as soil, water, air, and sunlight. Systems can be part of larger systems. For example, a large forest ecosystem may include a pond ecosystem. An ecosystem may be quite small, like a city park or an aquarium.

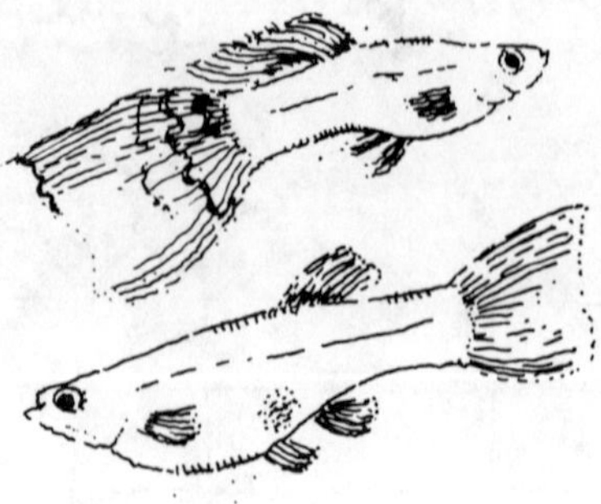

3. Explain that the class will conduct several experiments. The goal of these experiments will be to learn about how the fish interact with their ecosystem. The experiments will follow these steps:

a. Allow the fish **one day to adjust to changes** in their environment.

b. Make **maps** of where the fish swim in the aquarium, by observing and recording their locations every 30 seconds. This method of observing is called a ***sampling system.***

c. **Change one characteristic** in the aquarium, such as the number of fish, or the locations of plants. Then make a new map to see how the fish are influenced by the change in environment.

4. Divide the class into teams of five or six students, for recording the fish mapping experiments.

Making and Monitoring the Ecosystem

1. Ask the students for ideas on how the rocks in the aquarium could be rearranged to provide areas of light and shade, as well as hiding places for the fish. Encourage your students to discuss how different arrangements may affect the fish's environment. Have one team move the rocks to implement the class decision.

2. Have the students discuss the placement of the plants in the aquarium, and ask a second team to carry out the class decisions.

3. Point out that there are many differences in the areas of the aquarium. Ask students to describe some of these differences [lighter or darker, sheltered or open, bubbles or clear, plants or no plants, high or low oxygen levels, warmer or cooler, top or bottom].

4. Emphasize the importance of providing humane care for animals that are used for experimentation. To keep the fish healthy, student teams must measure and, if necessary, change the physical conditions of the aquarium.

5. Show the class the Aquarium Chart. Assign a weekly schedule for teams to measure the conditions in the aquarium. Explain and demonstrate how to make the various measurements. (See "Setting up an Aquarium," page 39, for details on each of these procedures.)

- **temperature:** Always read the same scale, in the same part of the tank, at about the same time of day. If the temperature goes below 60°F, or above 78°F, adjust the heater and again check the temperature every half hour or so until it stabilizes between 72°F and 78°F.

- **pH:** Demonstrate how to use a pH kit. Mix aquarium water with an indicator paper or solution (depending on the kit) and then compare the color of the indicator with a color chart. Since pH changes slightly during the day, always measure pH at the same time. The best range for most fish is between 6.8 and 7.0, with an acceptable range of 6.5 to 7.5.

- **water level:** Mark the level of the water on the first day with tape on the outside of the tank. It should be at least an inch (25 mm) below the top of the tank. Measure the distance between this mark and the actual water level. If the water level falls more than an inch (25 mm), add tap water that has stood at least overnight.

- **algae growth:** Watch for buildup of green material on the glass walls of the tank and the rocks. Estimate the algae buildup with a number from 0 (no algae) to 5 (thick green scum). Algae growth can be reduced by adding algae-eating fish or reducing the light level.

6. Designate two or three feeding times each day that do not overlap the time when students will be mapping fish movements. Tell the students to observe the fish closely to see if they eat plants and algae in addition to the food that is added to the tank. Remind students of the value of keeping a careful record of their observations.

7. Caution the class that overfeeding is one of the primary causes of death for aquarium fish, because the decaying food uses up oxygen in the water. Demonstrate how much food to put in the tank. [One pinch of flake food is sufficient.]

Getting to Know the Fish

1. Give the students about five minutes to share and discuss their prior knowledge of aquarium fish.

2. If you are just setting up your aquarium, you may wish to have your students transfer fish into the aquarium, one at a time. If you are adding new fish to an existing aquarium, you may also wish to have the students assist.

3. If your aquarium is already established and you do not plan to add additional fish, assign teams to watch the behavior of the resident fish after the students have changed the aquarium environment.

4. Explain that each team will watch the fish for three minutes and then return to their seats to write down their observations while the next team releases and/or observes the fish.

5. Caution students to be quiet and "sneaky" in observing the fish. The goal of the observations is to find out how the fish initially respond to their new environment and the other fish. Remind the students that systematic mapping of fish movements should not begin until after the fish have had one day to adjust to the new environment.

6. Give each team a copy of the "Fish Facts" information sheet that you have prepared. Have students read and discuss the information as they await their team's turn.

7. Challenge the students to "get to know" their fish by looking for characteristics that distinguish it from the other fish.

8. After they have watched the fish and returned to their seats, give each student a piece of paper on which to write observations of fish behavior.

9. Save a few minutes at the end of the session for the teams to share and discuss their observations.

10. Encourage teams to make informal observations of the fish and monitor the aquarium environment, while the fish get adjusted to the new environment.

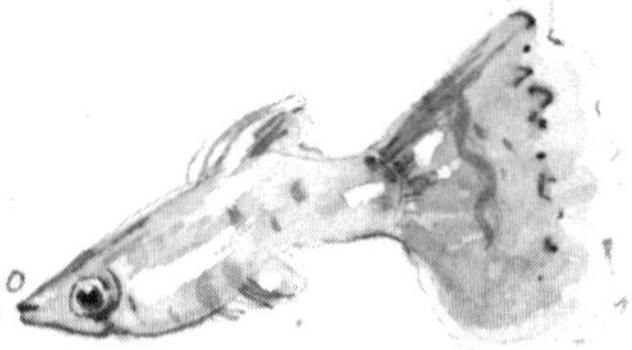

Fish Facts

GUPPY ▲ *Poecilia reticulata*

ROSY BARB ▲ *Barbus conchonius*

BRONZE CATFISH ▲ *Corydoras aeneas*

Native to South America, guppies eat mosquito larvae and other insects at the surface of the water. Standard guppy flake food contains mosquito larvae, brine shrimp, other protein foods, and plants. Mated pairs of guppies breed frequently, and the female gives birth to live young. The male guppy's anal fin is rod-shaped and the female's is fan-shaped.

Native to northeastern India, rosy barbs travel together in schools. They can be fed on flake food or tablet food and will also nibble plants. The males have more red and black on their fins than the females, and turn a deep red color at breeding times. They breed frequently and will scatter their eggs around the aquarium.

Native to South America, bronze catfish travel in schools. They can be fed flake food or tablet food, but will also scavenge for food along the bottom and eat algae and plants. When breeding, the females will often place the fertilized eggs on plants.

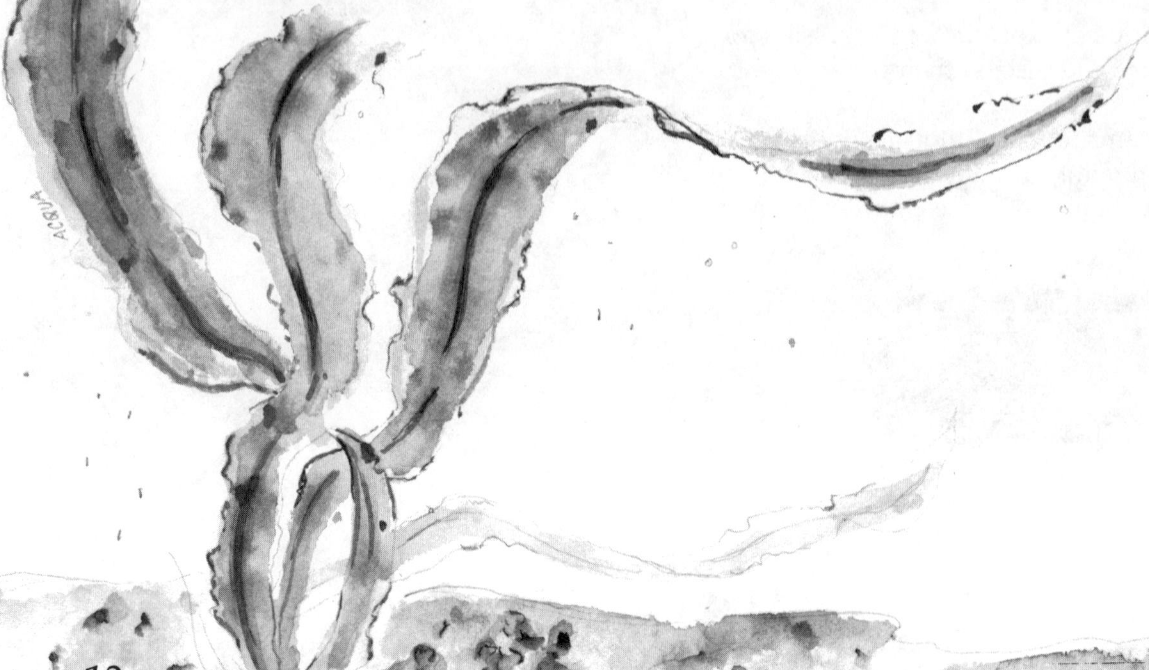

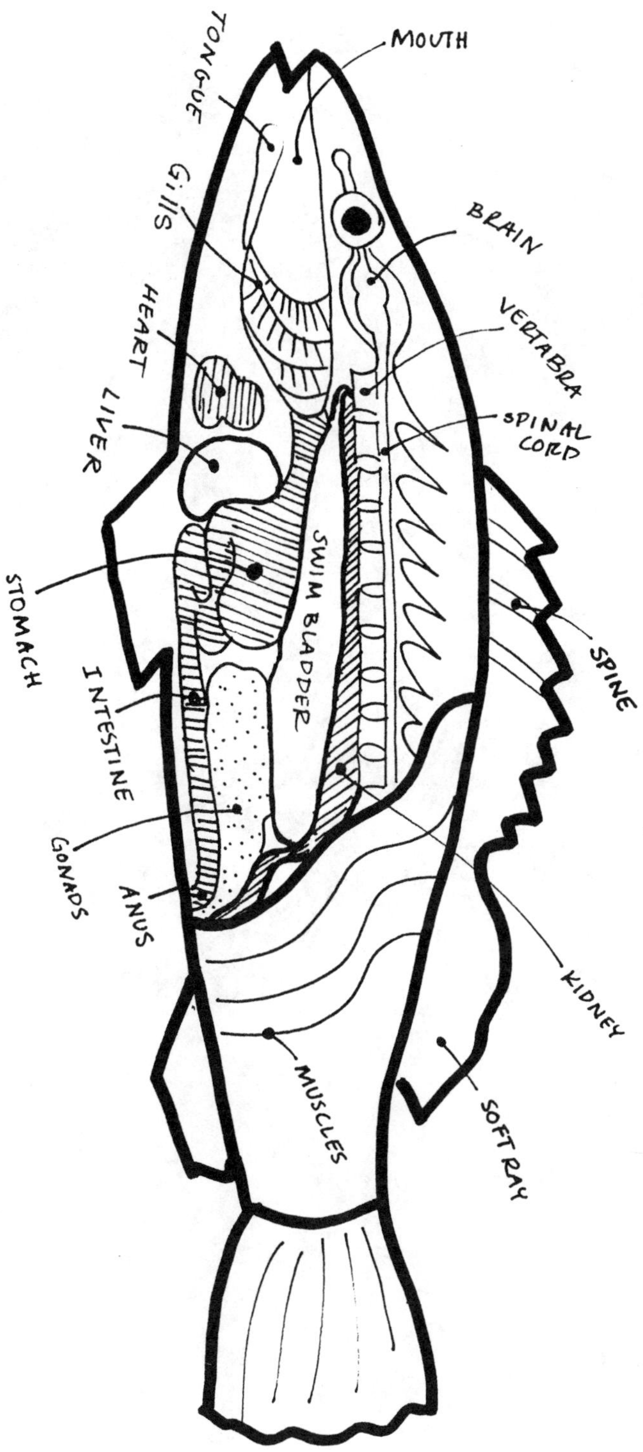
TONGUE
MOUTH
GILLS
BRAIN
HEART
VERTABRA
LIVER
SPINAL CORD
STOMACH
SWIM BLADDER
SPINE
INTESTINE
GONADS
ANUS
KIDNEY
MUSCLES
SOFT RAY

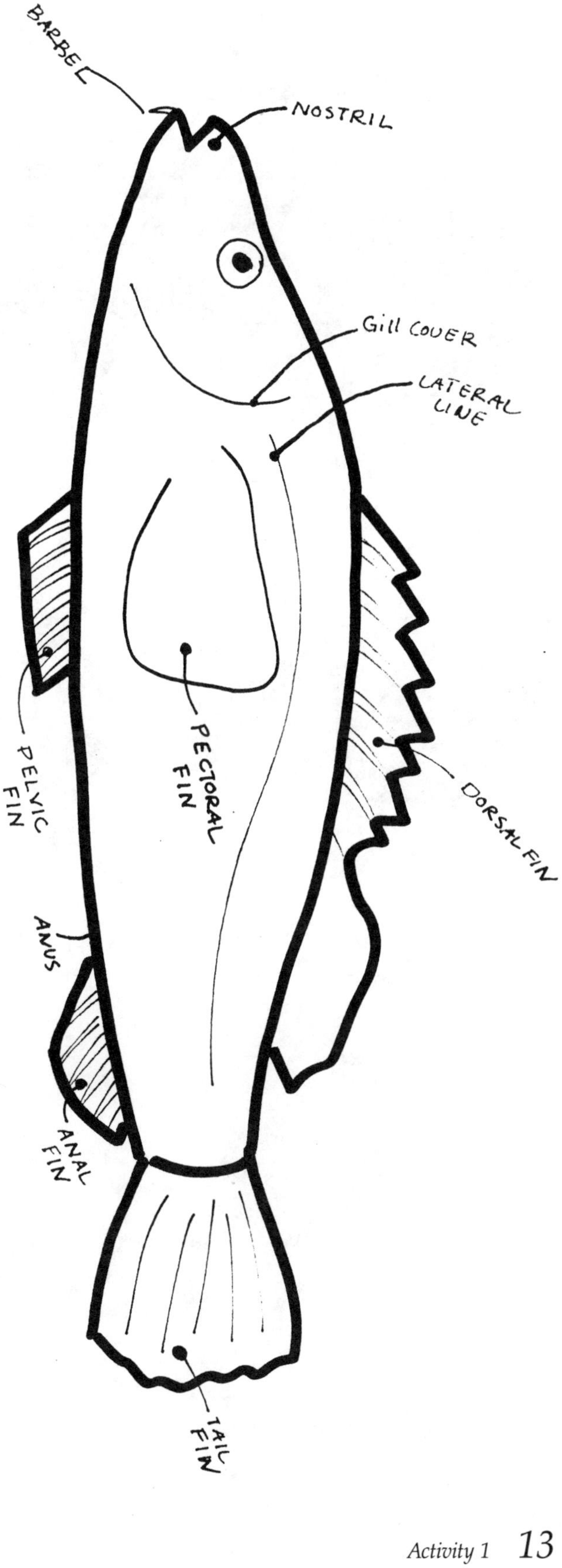
BARBEL
NOSTRIL
Gill COVER
LATERAL LINE
PELVIC FIN
PECTORAL FIN
DORSAL FIN
ANUS
ANAL FIN
TAIL FIN

FISH HABITATS

Activity 2: Mapping Fish Movements

Overview

In this activity, the students compare the different fish so that they will be able to identify each individual during the mapping activities. They are then asked for their observations of what each species of fish needs to survive. The purpose of this part of the activity is to stimulate the students' thinking about fish survival needs, and to introduce the concept of ***habitat.*** Teams then learn the procedure for mapping fish movements, and the first team completes its mapping activity.

Time Frame

Discussing Habitat Requirements — 10 minutes
Introducing the Fish Mapping System — 10 minutes
Demonstrating the Fish Mapping System — 10 minutes

What You Need

For the class:

- ☐ 1 grease pencil for drawing on acetate
- ☐ 1 roll of clear tape
- ☐ 1 piece of cardboard (slightly larger than the front wall of the aquarium)
- ☐ 1 or more sheets of white paper to cover the cardboard

For each team of 5–6 students:

- ☐ 1 sheet of clear acetate or plastic, cut to fit the front wall of the aquarium
- ☐ 50 or 60 half-inch adhesive dots, 10 of each color (one color for each fish)
- ☐ 1 manila folder

For each student:

- ☐ 1 "Fish Observations" student sheet (master included on page 21)

Name

Fish Observations

Date	Kind of Fish	Describe the Actions of the Fish

Getting Ready

1. Trim a piece of cardboard so it is slightly larger than the front wall of the aquarium. Cover the cardboard with white paper to provide a bright background, and draw an outline that matches the front wall of the aquarium. (This data board will be used by each successive team to map fish movements.)

2. For each team, cut one sheet of acetate the same size as the front wall of the aquarium. Tape one acetate sheet over the aquarium outline on the cardboard.

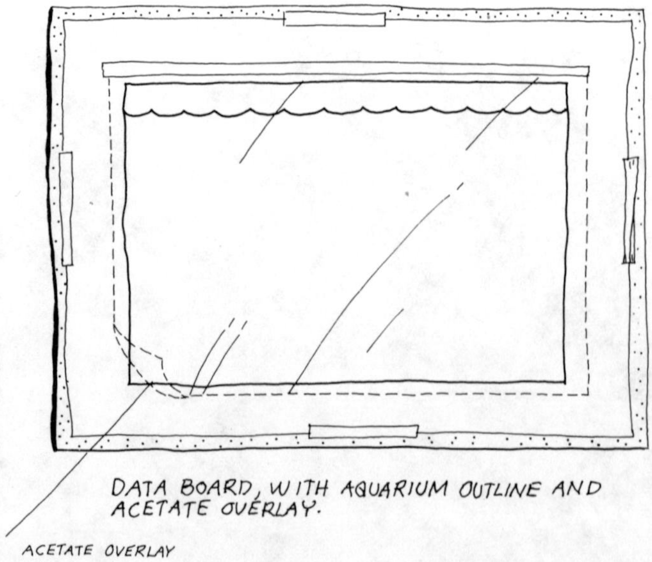

DATA BOARD, WITH AQUARIUM OUTLINE AND ACETATE OVERLAY.

3. Duplicate one copy of the "Fish Observations" sheet for each student.

4. At the top of the chalkboard, write the heading: "WHAT FISH NEED TO SURVIVE." Write the names of the different species of fish as row headings down the left side of the chalkboard. Below each name, sketch an outline of the fish, so that it can be distinguished from the other kinds of fish. Across the top of the chalkboard, just under the heading, write the following categories: "Food," "Shelter," "Temperature," "Space," and "Special Needs."

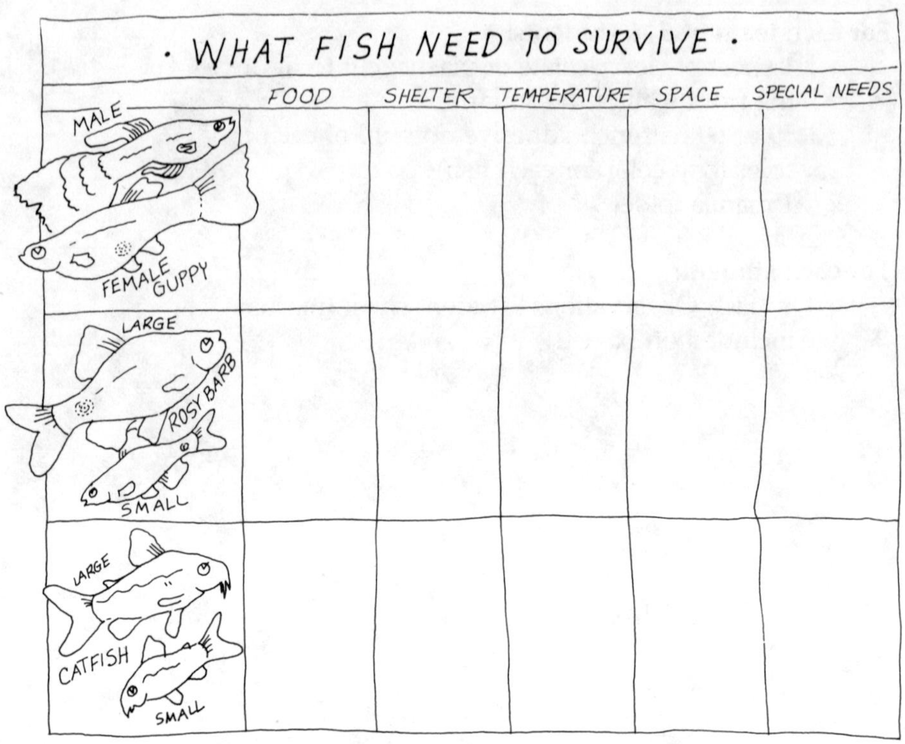

Discussing Habitat Requirements

1. Review the names of the different kinds of fish drawn on the chalkboard, and discuss the characteristics used in telling them apart.

2. If you are using two fish of the same species, ask students to describe characteristics that will identify each individual. Draw a second sketch on the chalkboard so each fish is represented.

3. Tell your students they may refer to their "Fish Facts" information sheet during the following class discussion. Encourage students to share additional knowledge they may have from any prior experience in observing or taking care of fish.

4. Starting with the category "Food," ask your students what each fish eats. (If your students reply "fish food," ask a student to read the list of ingredients on the food container. At this stage, students may not realize that some fish eat algae and other plants in addition to the fish food.) Write student responses on the chalkboard chart.

5. Ask your students about their observations of other fish needs, and fill in the chart for "Shelter," "Temperature," "Space," and "Special Needs." For the "Space" category, ask students to predict the area of the tank in which they think each fish will tend to spend the most time. Encourage them to make inferences based on the body shapes of the fish and their observations so far. The last category, "Special Needs," is open-ended, to allow students to talk about anything they might have read about or noticed during their informal observations, such as places to breed.

6. Define the word ***habitat*** as the area where an animal or plant lives. Within the boundaries of its habitat the organism can find everything that it needs to survive and successfully reproduce during the course of its entire life cycle. Point out that each row on the chart is a brief description of each animal's habitat.

You may want to discuss the differences between an ecosystem and a habitat. In general, an ecosystem is all of the living and non-living things in a given area, viewed as parts of a system with many interactions. A habitat is the area in which a specific organism lives that contains all that the organism needs to live. People define the boundaries of ecosystems somewhat arbitrarily, depending on what they perceive to be parts of the system. A habitat can include parts of several ecosystems and must include all that the organism needs during its entire life cycle.

Introducing the Fish Mapping System

1. Point out to your students that so far they have learned a lot about the fish just by watching them informally. However, they can learn a lot more about fish behavior by observing and recording fish movements systematically.

2. Display the different colors of dots. Have students decide which color of dot will represent **each fish.** Ask a volunteer to make a "Fish Key" on a piece of paper to show which dot represents which fish. (See drawing.) Post the "Fish Key" near the aquarium.

3. Show the class the data board with its outline of the aquarium and acetate sheet. Explain that each team will receive a sheet of acetate on which to map the movements of the fish. One person on the team will use the grease pencil to label the acetate and trace an outline of the aquarium, rocks, plants, and surface of the water.

4. Explain that at the same time each day, one team at a time will map the locations of all of the fish. When the teams have finished their observations, the transparent maps made by the different teams will be overlaid and compared.

5. Hand out a "Fish Observations" sheet to each student. Explain that each person will keep a log of behaviors and conditions that do not show up on the maps.

6. Ask students to suggest examples of behaviors that might be described on the "Fish Observations" sheet (such as fish bumping or biting each other, speed of swimming, fin and mouth movements, feeding, breeding, or injuries such as torn fins).

7. Explain that before beginning the mapping activities, each team should spend two minutes quietly observing the fish. Then the team should spend five minutes mapping movements, followed by several minutes of writing down their observations on the "Fish Observations" sheet.

8. Designate a sequence in which the teams will map the fish movements over the next few days, and post the schedule. Select a standard time for observations, such as 15 minutes before the end of the science period.

Demonstrating the Fish Mapping System

1. Guide the first team through the sampling system as a class demonstration.

   a. Assign one team member to each fish. Emphasize that the team must cooperate and be well organized in order to observe the movements of all the fish at the same time. Observing all fish simultaneously may reveal interactions among the fish.

   b. Give one member of the team a grease pencil and piece of cardboard with acetate. Ask this student to trace the location of the rocks, plants, and surface of the water. Have the student draw from a position that is "eye level" with the center of the tank.

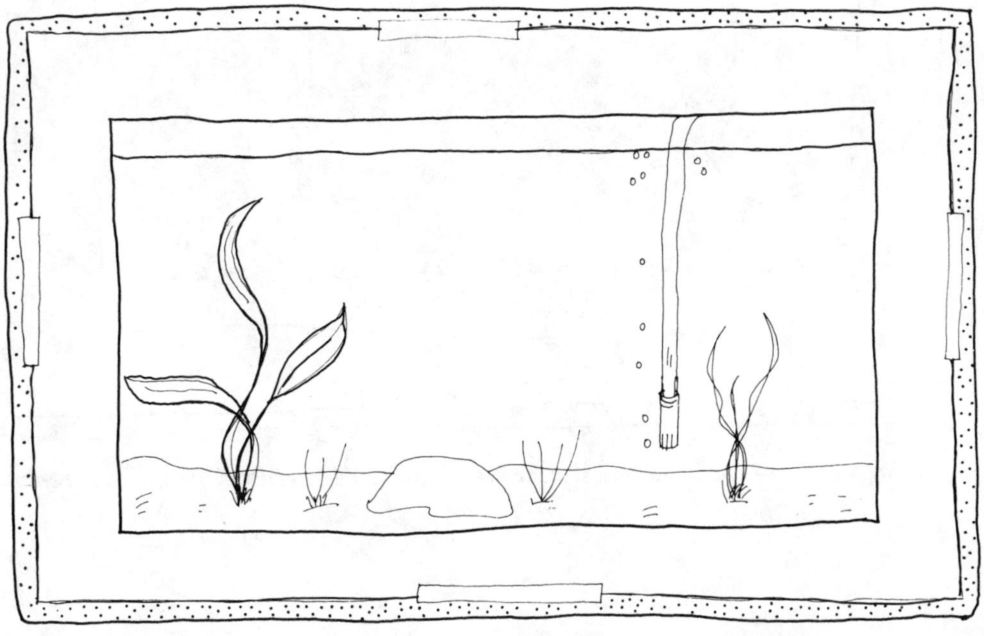

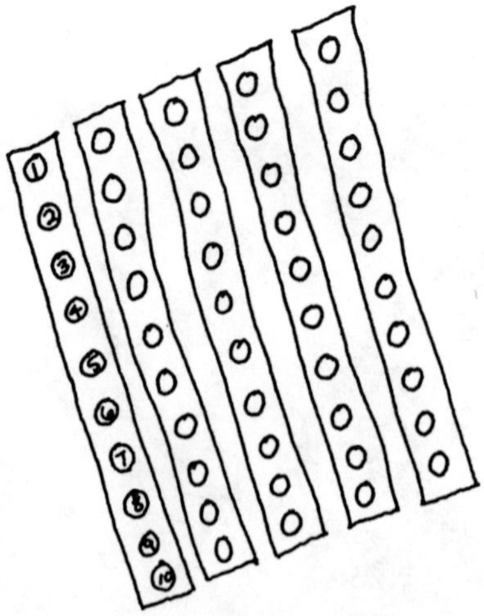

c. While the student is drawing, have the other team members number (from 1 to 10) the colored dots for each fish.

d. Appoint a Timekeeper to say "Time" every 30 seconds for five minutes.

e. Have every team member (including the Timekeeper) observe a different fish, and put a dot representing that fish on the map every 30 seconds, until 10 dots have been placed on the map for every fish.

2. Ask a team member to hold up the data board, so that the class can see the results of the first observations.

3. Answer any questions about the observation system and put the data board aside for analysis during the next activity.

If your students have difficulty keeping track of all the fish at once, an alternate method is to map the positions of half of the fish for five minutes then map the remaining fish for five minutes. This method allows students to assist each other in placing the dots, and frees the timekeeper from mapping responsibilities, but is not as accurate in recording the interactions among the fish.

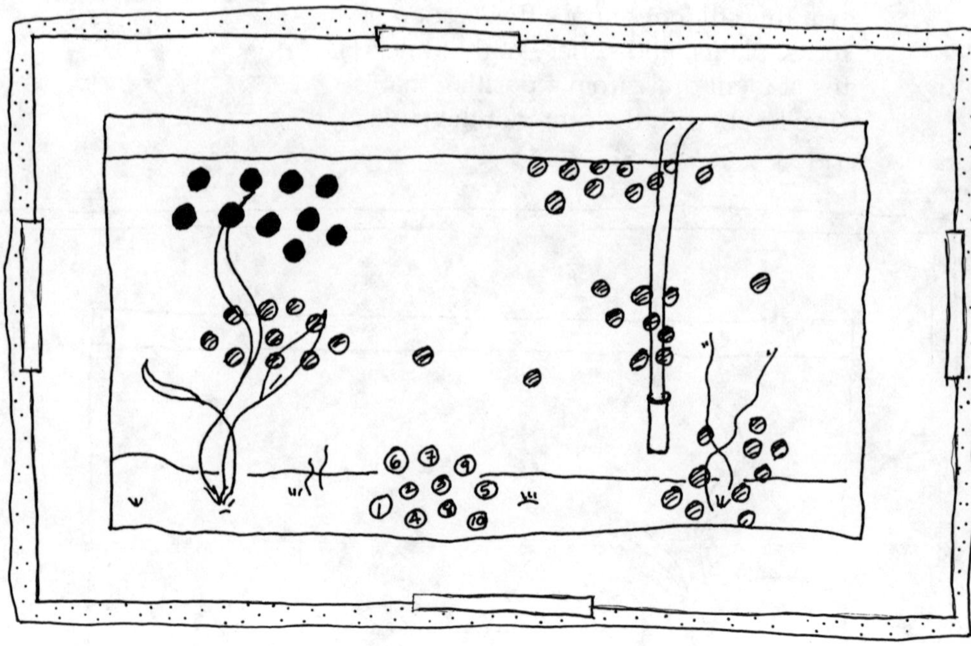

Placing the colored dots in numerical order (1–10) shows the sequence of the fish movements.

Name ______________________

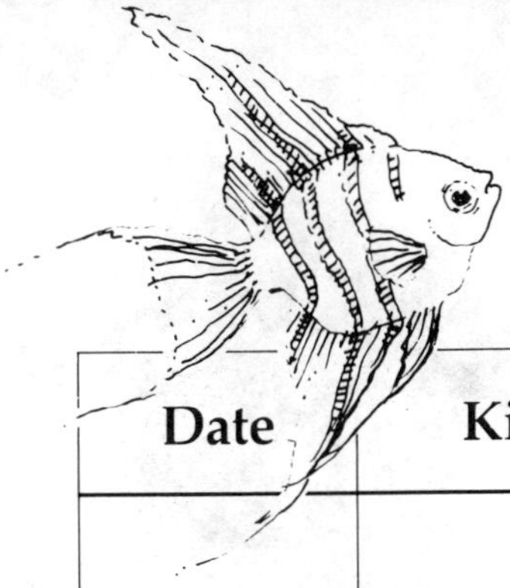

Fish Observations

Date	Kind of Fish	Describe the Actions of the Fish

Activity 3: Finding the Home Range

Overview

An animal's ***home range*** (area where it travels) is a clue to the size of the habitat that the animal needs to survive and successfully reproduce. In this activity, the class analyzes the map created by the first team to find the home ranges of the various fish. The students learn that biologists use similar ***sampling systems*** to study fish in their natural habitats.

The purpose of the mapping activity is to systematically record the home range of each fish. Once the students know how to do this, they are ready to start experimenting, or changing the environment in one way, and making a new map of fish movements in order to determine how the change affects the home ranges.

Time Frame

Analyzing the Data	20 minutes
Conducting an Experiment	20 minutes

What You Need

For the class:

- ☐ 1 data board with team map from the previous activity
- ☐ objects for experiments, such as: rocks, plants, wood, additional fish, a mirror, black paper, fluorescent lamp
- ☐ 1 grease pencil
- ☐ 1 roll of clear tape

For each team of 5–6 students:

- ☐ 1 sheet of clear acetate
- ☐ 5–6 colors of half-inch adhesive dots, 10 of each color (one color for each fish)
- ☐ team folder with "Fish Observations" sheets for each student
- ☐ chalkboard and chalk, or butcher paper and marker

Analyzing the Data

1. Display the data board with the map made by the first team so that all of the students can see it. Ask volunteers to describe the movement patterns of each of the fish. Ask the members of team 1 to refer to their "Fish Observations" sheets for records of other interesting behaviors.

2. Introduce the concept of ***home range*** as the area in which an animal travels during normal activities. Demonstrate how to find the home range of one of the fish by drawing a line on the map with a grease pencil to connect the outermost dots for one fish. The area outlined, or home range, will be a closed shape that includes all of the locations where the fish traveled while it was being observed.

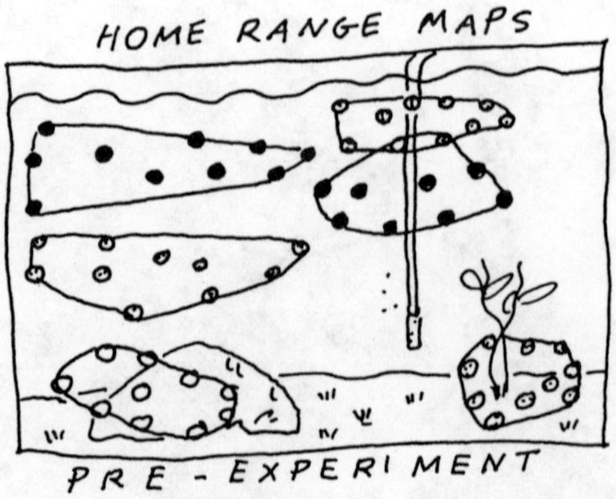

3. Ask for volunteers to come up to the data board and draw the home ranges for each of the other fish.

4. Explain that ***home range maps*** are used by biologists to estimate the size of the habitat needed by an animal to survive and successfully reproduce. The method of making the map by observing and recording the location of the animals at regular intervals of time is called ***sampling.***

Sampling System: *a method of observing and recording data at certain time intervals. In the case of behavior, which occurs continuously throughout the animal's life, the observer records an animal's location or describes what it is doing at times that have been selected in advance.*

5. Ask the students what they can infer about the habitat needs of each of the fish by looking at the home range map.

6. Ask the students to describe some relationships between the physical structures of the fish and their habitat needs. For example, the students might notice (or you might point out) that:

   a. The guppy's mouth opens slanting upward; its body is flattened or streamlined vertically, suggesting that it feeds near the surface.

   b. Catfish have mouths with "feelers" that open downward and their bodies are flattened horizontally, so they are well adapted for gathering food from the bottom.

Conducting an Experiment

1. Direct the students' attention to the materials you have assembled for the experiments. Ask for ideas about how each of these materials might be used to change the environment in a way that might affect the parts of the aquarium used by the fish. List each idea on the chalkboard or a large sheet of butcher paper. Plan to keep the list posted for a week or so, until all of the experiments are completed.

2. As each idea for changing the environment is listed, ask the students how they think the fish's behavior might change. For example:

- Put mirror at one end

"They might fight the 'other fish' in the mirror."
"Some fish will stay away from the mirror."
"No difference. They won't know what it is."

- Move rock hiding places to other side of tank

"Fish that hide will move home range to other side."
"No difference—they aren't hiding!"

3. Have the class vote on **one change** to make in the aquarium environment. Emphasize that if they make two or more changes, and the animals do change their home ranges or other behaviors, we will not know which change caused the response. So, it is important to make just one change for each experiment.

4. Have students in the second team change the aquarium in the way that the class agreed. Meanwhile, ask the rest of the class to make predictions about how each of the different fish will react to the change.

5. Explain that the second team will make a new home range map on the following day, after the fish have had time to adjust to the change in their environment.

Activity 4: Daily Experiments in the Aquarium

Overview

In this activity, the second team observes and maps the fish, then the class compares the maps made before and after the first experiment to determine how the change affected the home ranges of the fish. The students then infer habitat requirements based on the results of their experiment. The students discuss observations of other fish behaviors they have observed, and the teacher introduces the concept of *animal territories*.

The class then selects the next experiment from the list of student ideas for ways to change the aquarium ecosystem. The third team conducts the experiment. Experiments continue, one each day, until all of the teams have constructed a home range map.

Time Frame

Mapping Fish Movements	10 minutes
Discussing the Results	20 minutes
Conducting the Next Experiment	5 minutes
Daily Experiments Continue	30 minutes/day

What You Need

For the class:

- ☐ 1 data board with team map from the previous activity
- ☐ objects for experiments, such as: rocks, plants, wood, additional fish, a mirror, black paper, fluorescent lamp
- ☐ 1 grease pencil
- ☐ 1 roll of clear tape

For each team of 5-6 students:

- ☐ 1 sheet of clear acetate
- ☐ 5–6 colors of half-inch adhesive dots, 10 of each color (one color for each fish)
- ☐ team folder with "Fish Observations" sheets

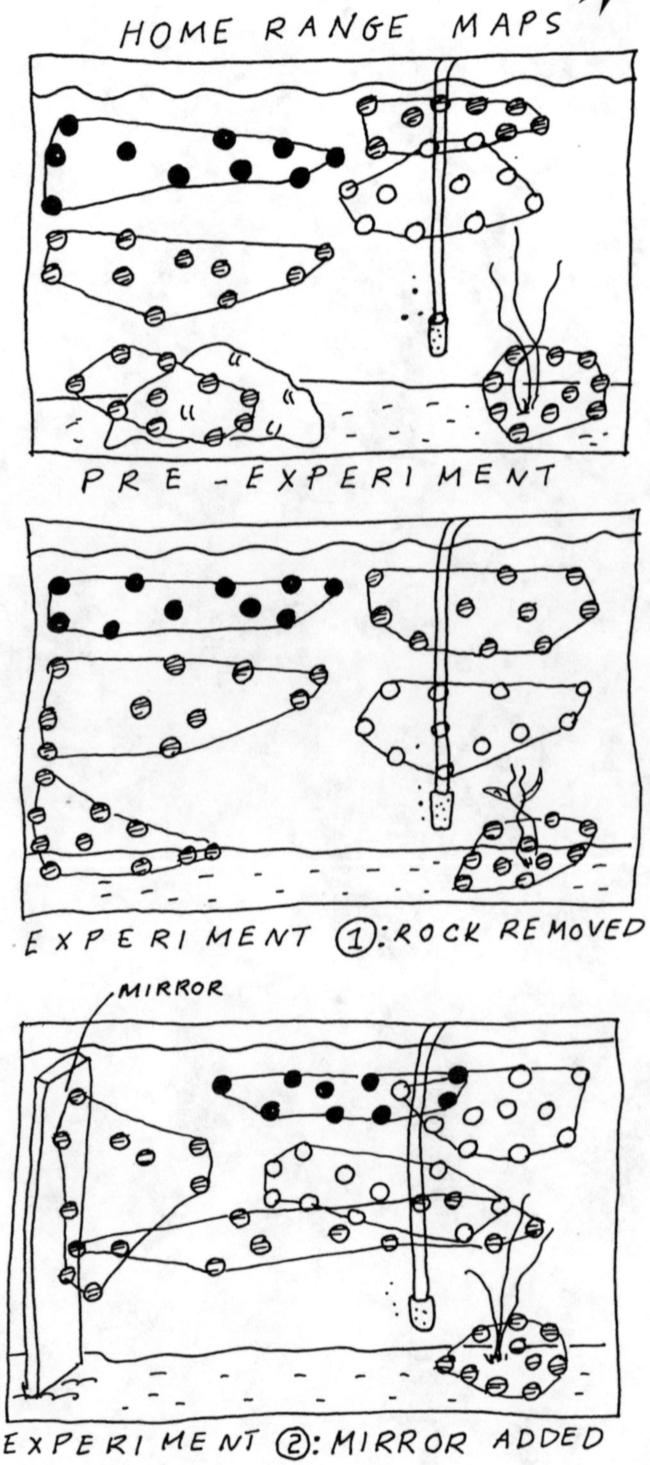

Mapping Fish Movements

1. Tell the second team to:

   a. Observe the fish informally for two minutes, and record observations on their "Fish Observation" sheets.

   b. Make a new home range map.

   c. Use the grease pencil to draw in the home ranges for each fish.

2. You may wish to have the class go on to other work while the second team is busy observing and mapping. When the second team finishes, tape the "pre-experiment map" (made by Team 1 during Activity 3) next to the "Experiment 1" map made by Team 2. Post the maps on the wall or bulletin board so the students can see them when discussing the results during the next activity.

Discussing the Results

1. Ask the students what they predicted might happen when the environment of the aquarium was changed. Point out the "pre-experiment" and "Experiment 1" maps on the wall. How did the predictions compare with the mapping records?

2. To compare the movement patterns and home ranges of the various fish, remove the "Experiment 1" map from the wall, and tape it over the "pre-experiment" map. Ask the students to identify changes in movement patterns, as well as patterns that didn't seem to change. Ask for any additional behaviors that were recorded on the "Fish Observation" sheets.

3. Ask the students what they can infer about the fish's habitat needs from the experiment.

4. Tell the students that some kinds of animals defend an area of their habitat against other animals. This area is called a ***territory.*** Ask the students:

- "Have you seen any behavior that looked like a fish was defending its territory?"
- "Can you suggest reasons why animals might defend territories?" [protect mates, young, food, shelter, etc.]
- "Can you name some other kinds of animals that defend a territory?" [humans, wolves, cats, many species of birds and fish.]

Conducting the Next Experiment

1. Refer to the list of ideas about ways to change the aquarium. Invite students to give their opinions about what experiment they would like to try next. For each idea, ask students to predict what the results might be.

2. Have the class vote on one way to change the environment for the second experiment. Ask for opinions about whether to keep the change introduced by the first experiment, or to return the aquarium to its original state. (Either decision is okay, since there is a record of home ranges with or without the change. If the class decides to keep the change introduced in the first experiment, they should use the map made after the first experiment for comparison. If they remove the change, they should use the "pre-experiment" map for comparison.)

3. Assign the third team to change the aquarium in the way that the class agreed. Meanwhile, ask the rest of the class to make predictions about how each of the different fish will react to the change.

Daily Experiments Continue

1. Tell the third team to:

   a. Observe the fish informally for two minutes, and record observations on their "Fish Observation" sheets.

b. Make a new home range map.

c. Use the grease pencil to draw in the home ranges for each fish.

2. As before, you may wish to assign other work for the rest of the class as the third team conducts its experiment. Post the results for the students to see.

3. Set aside about 15 minutes of each session for the class to discuss the results of the previous experiment. You may wish to guide the discussion with questions like the following:

- What did you predict?
- What are the differences between the two mapping records?
- What do the results of the experiment tell us about the fish's habitat needs?
- What other behaviors did you observe?

4. After discussing the previous experiment, ask for ideas for the next experiment. The list of possible changes in the aquarium may be used as a source of ideas, but should not limit the possibilities. Students may suggest bringing other materials or equipment to school to test their hypotheses. Encourage creativity and cooperation in deciding what experiment to perform next.

5. Have students continue to monitor and regulate the environmental characteristics of the aquarium.

6. Continue to plan and conduct experiments until each team has had an opportunity to observe and map the fish movements. You may wish to extend the experiments as long as you and your students find the activity challenging and can think of new variables to change. Small groups of students may wish to continue experiments outside of regular class time as a special project.

Going Further

1. Introduce the terms ***dominant*** and ***subordinate*** and challenge the students to find out which fish is "boss" (the most dominant of all). You might ask the students whether or not they think a more dominant fish would grow more rapidly than the others. (Dominant fish tend to get more of the food, so they often do grow larger.) Suggest that students look at the relative growth rates of the different fish for evidence of a ***dominance heirarchy***.

2. To allow the fish to display their predatory behaviors, occasionally have the students feed the fish live organisms, such as tubifex worms, daphnia, brine shrimp, or small insects.

Encourage your students to apply their mathematical abilities to create a graph analyzing population growth over time.

3. Follow the life cycle of a few aquatic snails to provide an example of an ***herbivore*** (plant eater) in the aquatic ecosystem. Snails reproduce rapidly, so start with a small number. Have your students make regular population counts. Suggest that they graph the results of their "snail census" over a period of several months.

4. If you haven't done so already, introduce one male and two female guppies into the aquarium. Have students monitor the changes in the size of the population over several months. Encourage students to explore the concept of **carrying capacity**.

5. Use the aquarium to study the interactions within an ecosystem. Have students identify the things produced within the aquarium and the things that travel between the aquarium and the classroom (light, air, food, etc.). You may want to explore the oxygen, carbon dioxide, and nitrogen cycles.

6. Purchase a variety of fresh fish and have students investigate the external adaptations and internal structures. Rock cod, perch, and flounder make good subjects for fish prints. If you use non-toxic paints to make the prints, the fish can be washed then cooked for a gourmet "fish-tasting" extension.

Summary Outlines

Activity 1: Setting up the Fish Tank

Getting Ready

1. Set up aquarium.
2. Select 5 or 6 fish of three different species.
3. Prepare "Fish Facts" information sheet.
4. Make an Aquarium Chart.
5. Make a Daily Feeding Schedule.
6. Decide on rocks and plants that students can change.
7. Purchase fish close to first day of class.

Introducing the Project

1. Tell students that they will form teams to study fish behavior.
2. Define ecosystem. The aquarium is an ecosystem.
3. Goal is to learn how fish interact with ecosystem and each other. In each experiment:
    a. Allow fish one day to adjust to changes.
    b. Map where fish swim; record location (observe continuously) every 30 seconds.
    c. Change one characteristic of environment.
4. Divide class into teams of 5-6 students.

Making and Monitoring the Ecosystem

1. Discuss how to rearrange the rocks in the aquarium.

2. Discuss how to arrange the plants in the aquarium.

3. Students describe different areas of aquarium.

4. Emphasize humane care of fish.

5. Show Aquarium Chart and discuss:

   temperature—good range is 72° to 78°F.

   pH—good range is 6.5 to 7.5.

   water level—add water if the level falls more than 1″.

   algae growth—estimate algae buildup with a number from 0 to 5.

6. Designate feeding schedule; observe feeding behaviors.

7. Caution class not to overfeed.

Getting to Know the Fish

1. Discuss students' knowledge of fish.

2. If you are adding fish, have students assist.

3. If not adding fish, assign teams to observe resident fish.

4. Tell students they will have 3 minutes to observe fish.

5. Caution students to be "sneaky" in observing the fish.

6. Give each team a copy of the "Fish Facts" information sheet.

7. Challenge students to "get to know" their fish.

8. Give students paper on which to write fish observations.

9. Discuss observations.

10. Teams observe informally; fish get used to aquarium.

Activity 2: Mapping Fish Movements

Getting Ready

1. Make cardboard data board.

2. Cut 1 sheet of acetate for each team.

3. Duplicate 1 copy of "Fish Observations" sheet for each student.

4. Make chalkboard chart: What Fish Need to Survive.

Discussing Habitat Requirements

1. Discuss how to tell different species of fish apart.

2. Discuss how to identify individual fish.

3. Fill in chart on chalkboard under "Food" category.

4. Students discuss other fish needs to fill in rest of chart.

5. Define habitat. Distinguish from ecosystem.

Introducing the Fish Mapping System

1. Stress the importance of making systematic observations.

2. Decide which dot colors represent which fish. Make "Fish Key."

3. Show data board and grease pencil, and explain how to trace outline of aquarium.

4. Each day another team maps fish locations.

5. Hand out "Fish Observations" sheet.

6. Discuss different fish behaviors that might be listed.

7. Teams observe informally for 2 minutes, map for 5 minutes, then record on "Fish Observations" sheet.

8. Designate a time of day and sequence for team observations.

Demonstrating the Fish Mapping System

1. Guide first team through system as a demonstration:

    a. Assign one team member to each fish.

    b. One student outlines rocks, plants, etc.

    c. Other team members number dots.

    d. Appoint Timekeeper.

    e. Put one dot for each fish on chart every 30 seconds for 5 minutes.

2. Display the data board, discuss results, and answer questions.

Activity 3: Finding the Home Range

Analyzing the Data

1. Display the data board and discuss movement patterns.

2. Define home range and draw home range for one fish.

3. Ask volunteers to draw home ranges of other fish on data board map.

4. Explain that home range maps indicate what animals need in their habitat.

5. Ask students to infer habitat needs of each fish.

6. Ask students to describe relationships between physical structures of fish and their habitat needs.

Conducting an Experiment

1. List students' ideas for experiments.

2. Discuss how fish might respond to changes in aquarium environment.

3. Class votes on one change for first experiment.

4. First team changes aquarium; other students predict results.

Activity 4: Daily Experiments in the Aquarium

Mapping Fish Movements

1. Tell second team to:

   a. Observe fish informally for 2 minutes.

   b. Make new home range map.

   c. Draw in home ranges for each fish.

2. Team finishes map, and displays on wall.

Discussing the Results

1. Compare predictions with map records.

2. Overlay the two maps. Identify changes in home ranges.

3. Infer habitat needs from experimental results.

4. Define territory. Ask questions about animal territories.

Conducting the Next Experiment

1. Discuss ideas for next experiment.

2. Vote on one way to change the aquarium. Decide which maps should be compared.

3. Assign team to make change; ask class for predictions.

Daily Experiments Continue

1. Tell third team to:

    a. Observe informally for 2 minutes.

    b. Make a new home range map.

    c. Draw in home ranges with grease pencil.

2. Post results for students to see.

3. Discuss previous results.

4. Discuss ideas for the next experiment.

5. Students continue to monitor aquarium.

6. Plan and conduct at least one experiment per team.

Setting up an Aquarium

Following is a brief overview of the basic information that you will need to set up an aquarium for hardy tropical fish. However, if you have never set up an aquarium before, or if you are using equipment such as a filter system or heater that you have not used before, it is best to supplement this overview with an aquarium guide. Guides are commonly available in pet shops, and several excellent ones are listed in the "Resources" section on page 45.

What You Need

- ☐ 1 resource guide on the care of aquarium fish
- ☐ 1 aquarium tank, ten to twenty-gallon capacity
- ☐ 1 plastic or metal aquarium cover
- ☐ 1 air pump with aerator and filter
- ☐ gravel (about 2 lbs. for every gallon of water in your tank)
- ☐ 1 or 2 large rocks
- ☐ live water plants (such as Elodea, Vallisneria, or Egeria) or several plastic plants
- ☐ 1 fishbowl or gallon jar as a temporary container
- ☐ 1 small fish net
- ☐ 1 aquarium pH kit
- ☐ 1 thermometer (either alchohol or liquid crystal)
- ☐ 1 siphon tube for aquariums
- ☐ 1 sieve or dishpan for washing gravel
- ☐ 1 container of flake fish food
- ☐ 1 container of tablet food for the bottom feeding fish

Optional:

- ☐ 1 bottle of dechlorination liquid (or tablets if available)
- ☐ 1 heater
- ☐ 5–10 aquatic snails
- ☐ 1 test kit for nitrites
- ☐ live fish food, such as tubifex worms, brine shrimp, daphnia, or small insects

Selecting a Tank

A large (20-gallon) aquarium is easier to maintain as a balanced system and better for your students to observe and map fish. The minimum size that you should plan to use for these activities is a 10-gallon tank.

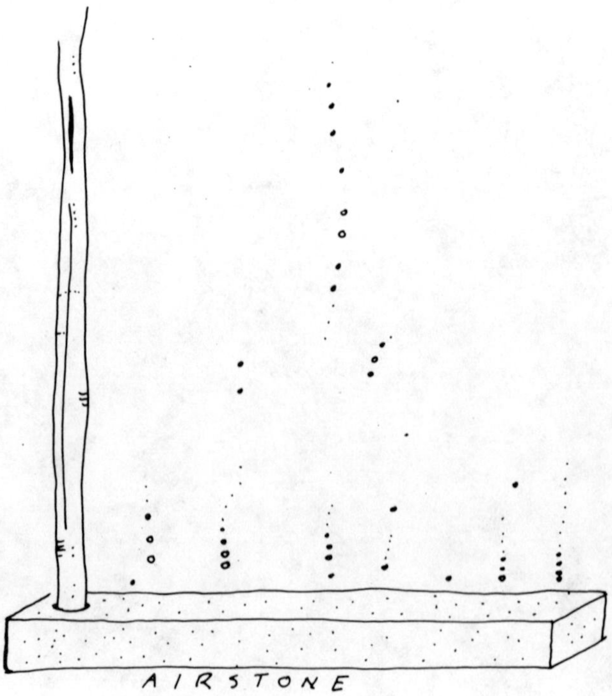

Selecting a Site

A 20-gallon tank filled with water weighs about 200 pounds, so it must be placed on a sturdy table or counter. Do not place it in direct sunlight because the temperature may vary greatly and too much algae will grow.

Dechlorinating the Water

In most areas, tap water is acceptable for use in the aquarium as long as it sits overnight to allow the chlorine to evaporate. However, if the water in your area is heavily chlorinated, or if it is high in copper or other metals, it is best to use bottled water, or to purchase dechlorination tablets. These tablets also remove metals and other minerals from the water.

Aeration

The gills of fish absorb oxygen dissolved in the water, and release carbon dioxide into the water. To provide a good oxygen supply at all levels in the tank, you will need an air pump attached with plastic tubing to an airstone that bubbles air through the water.

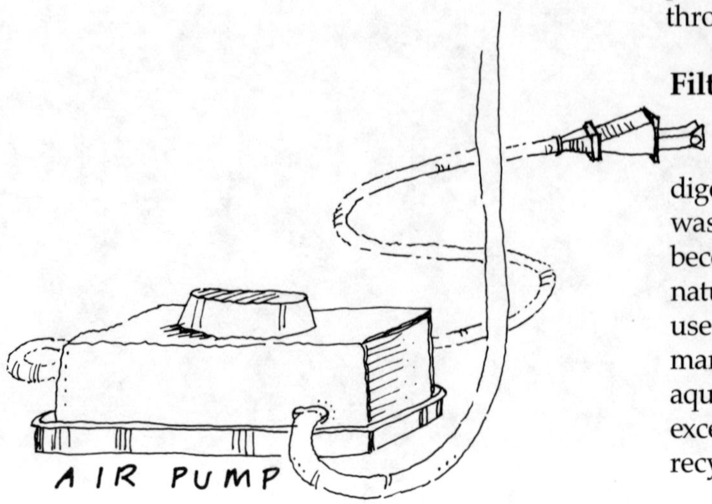

Filtration

In the process of eating and eliminating digested food, fish produce ammonia and nitrite waste products that accumulate in the water and become toxic to the fish. Plants and bacteria will naturally absorb these chemicals from the water to use in their own growth. However, if there are too many fish and/or too much decaying food in the aquarium, the ammonia and nitrite levels may exceed the abilities of the plants and bacteria to recycle the waste products.

An inexpensive foam cartridge filter can be added to the air line to help remove waste products from the water. The foam provides a surface for the growth of bacteria that absorb harmful chemicals from the circulating water. Alternatively, you may want to set up an "under gravel" filter system that circulates water and air through the entire gravel bed in order to promote the growth of helpful bacteria.

It is advisable to purchase a metal cover or make a plastic cover for your aquarium as this will reduce evaporation and heat loss, and will prevent fish from escaping.

Gravel, Rocks, and Logs

You will need about two pounds of gravel for each gallon of water. A medium-brown or tan color will help to camouflage the fish. Avoid fine grain sand. Also avoid very coarse gravel that will trap uneaten food. Wash the gravel thoroughly in plain water using a sieve or dishpan (do not use detergents).

Hard rocks, petrified wood, driftwood, bamboo cane, and synthetic logs provide fish with hiding places, without adversely affecting the water chemistry. Avoid new wood, coral, limestone, marble, and metallic rocks. These materials have harmful substances that dissolve, changing the chemistry of the water.

Plants

Artificial plants are inexpensive and will not die, decay, or be eaten by the fish. They provide hiding places and can be easily moved by your students during the course of their experiments.

Living plants cost more than artificial plants, are likely to be uprooted by bottom fish, and are a source of decaying material that may be harmful if not removed. On the other hand, living plants absorb toxic nitrites and minerals that may build up in the gravel. Plants also provide a source of food for some fish. Either artificial or living plants are acceptable for the activities in this Teacher's Guide.

Aquarium Cover and Light

It is advisable to purchase a metal cover or make a plastic cover for your aquarium as this will reduce evaporation and heat loss, and will prevent fish from escaping.

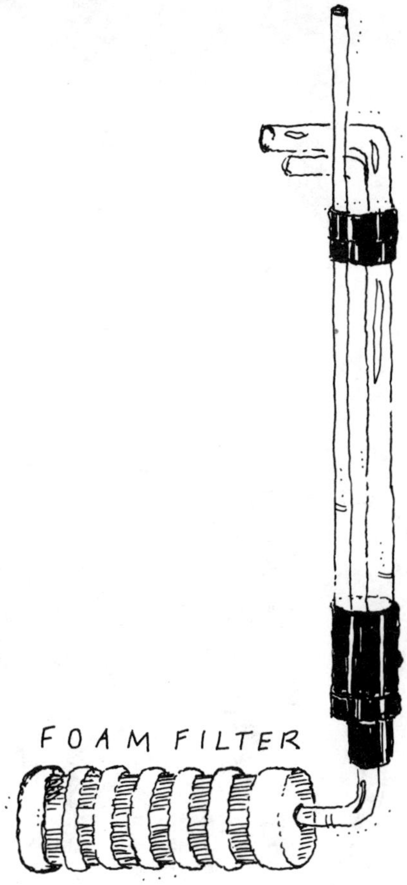

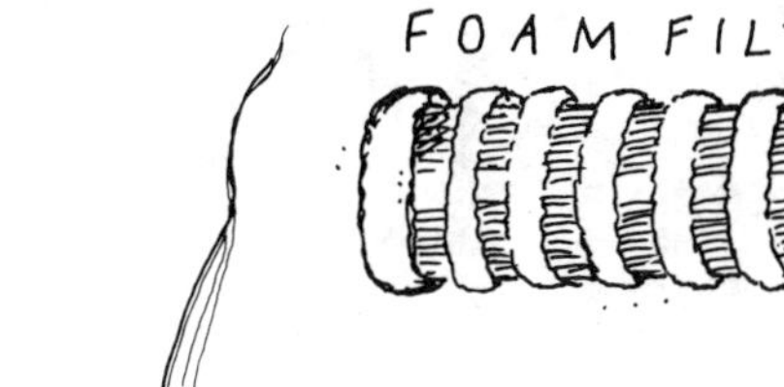
VALLISNERIA

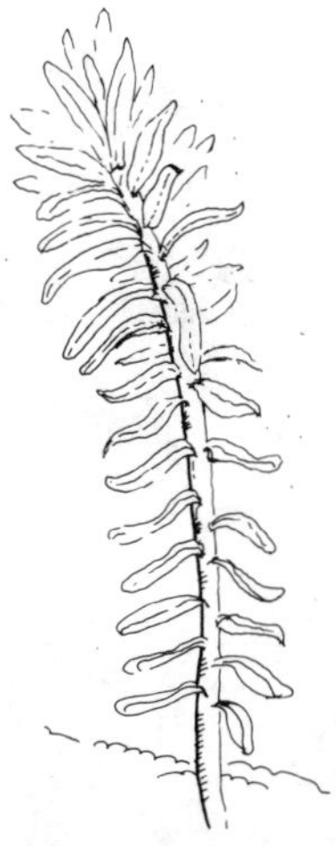
ELODEA

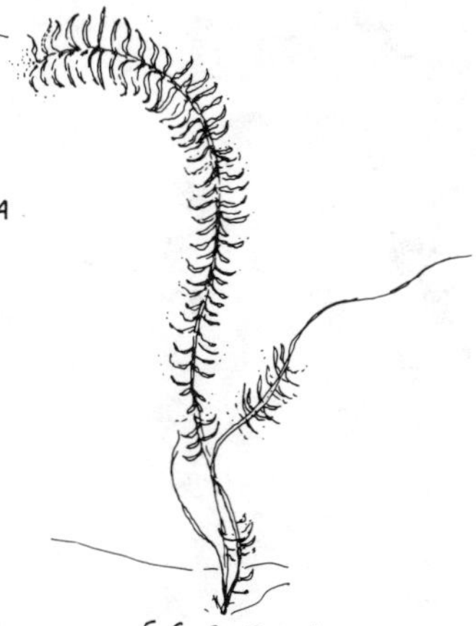
EGERIA

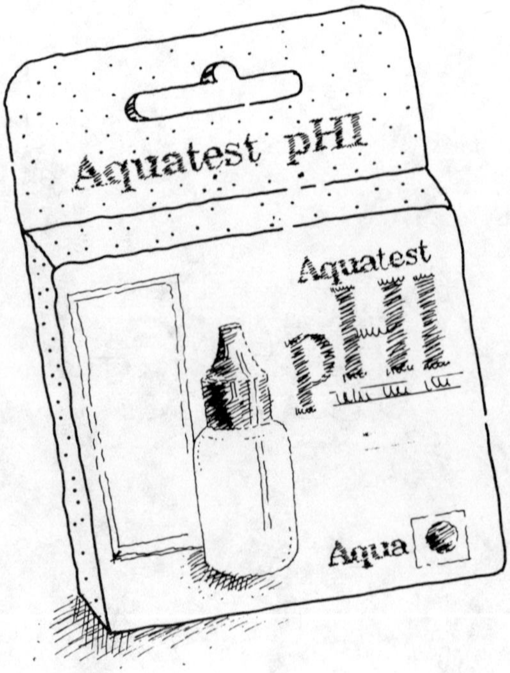

If your room is well-lighted and you use artificial plants, an aquarium light is unnecessary. However, if it is difficult to see the fish clearly, and/or you keep live plants in the aquarium, you will need to purchase an aquarium light. Fluorescent lighting is better than incandescent lighting as it will not overheat the aquarium.

Thermometer and Heater

Tropical fish will do best within a range of 72°F to 78°F (34°C–36.8°C), but most fish will survive if the aquarium does not drop below 60°F (28°C). If your classroom gets colder than that overnight, you will need a heater. A 20-gallon tank will require a heater rated at about 200 watts.

Whether or not you install a heater, you will need a thermometer so your students can monitor temperature. A liquid crystal thermometer taped to the outside wall of the aquarium is accurate and easy to read. You may also tape a standard alcohol thermometer to the inside corner of the aquarium so that its bulb is immersed in the water. Be careful not to locate the thermometer too close to the heater or light.

When you first set up the tank, check the water temperature daily. If you use a heater, be sure that the temperature does not go above 78°F (36.8°C).

Setting up the Aquarium

It is best to set up the aquarium **at least a week before adding the fish.** This time period will allow algae and bacteria to become established in the gravel, and will ensure that any chemical imbalances in the water have stabilized. If possible, arrange for a small group of students to assist you.

Arrange the gravel and large rocks on the bottom of the tank. Fill the aquarium three-quarters full. To avoid disturbing the gravel, pour the water onto a plate or your hand inside the tank. If you have decided to use dechlorination tablets, add them to the aquarium.

Arrange plants and smaller rocks or logs, and set up the airstone and filter system, and (optional) heater. Add additional water, leaving at least an inch of air space between the water surface and the cover.

Maintaining Water Quality

Several different pH kits are available in pet stores. They all require mixing water with an indicator paper or solution, and then comparing the color of the indicator with a color chart. Since the pH changes slightly during the day, always measure pH at the same time. The best range for most fish is between 6.8 and 7.0, with an acceptable range of 6.5 to 7.5. If the pH of tap water in your area falls outside of this range, plan to use bottled water.

The simplest way to maintain the acceptable pH range, while removing decayed food and other pollutants, is to replace about 20% of the water every three or four weeks. Use a siphon tube to bring up water from above the surface of the gravel. The replacement water should be the same temperature as the water in the aquarium, and should be dechlorinated by standing for a day, or by adding dechlorination tablets. Keep extra replacement water standing by to occasionally replace water that has evaporated.

Heavy growth of green algae on the walls and rocks of the aquarium may indicate an excess of nutrients in the water due to overfeeding. To control algae growth, add an algae-eating fish, reduce the light, and check that excess food is not accumulating in the tank. Algae deposits can also be rubbed off the inside of the glass with a magnetized scrubber that moves along the glass as you guide the movement with a magnet on the outside of the tank.

Ammonia and nitrites are the most common pollutants that can cause problems early in the life of the aquarium ecosystem, before the bacteria that normally absorb these chemicals have had a chance to become established. These pollutants tend to peak one to two weeks after the fish are first introduced. If the fish are not eating, have clamped fins, and become sluggish, replace 20% of the water in the aquarium using a siphon tube as described above.

You may want to purchase a water test kit for nitrites, as this is another interesting test that your students can do in their study of the aquarium ecosystem. This activity would be especially appropriate if your textbook discusses the nitrogen cycle.

Adding the Fish

After the aquarium has been left standing for at least a week, and the temperature and other conditions are in the correct ranges, you are ready to add the fish. It is recommended that you allow groups of students to do this during Activity 1, but this is not essential for the mapping activities.

When adding fish to the tank, be certain that the water temperature of the container is about the same as the temperature in the aquarium. It is advisable to "float" them in a plastic cup or bag for 10–20 minutes, so the container temperature gradually reaches the temperature of the aquarium. You can then pour the container water with the fish into the aquarium, or use a small net to transfer the fish.

Feeding the Fish

Your pet store operator can recommend a brand of flake food that contains a combination of ingredients that provide the fish with all of the nutrients they need. We also recommend that you purchase a tablet or pellet food for bottom feeders such as catfish. Live foods, such as tubifex worms, daphnia, or tiny insects are also commonly available from pet stores. Designate two or three feeding times during the day. It is better to give the fish small amounts of food two or three times a day than a lot of food once a day.

Resources

There are many excellent guides for setting up and maintaining an aquarium. These are available at most libraries and pet stores. Following are a few that we have found useful.

Baensch, U. and Baensch, H.A. *Digest for the Successful Aquarium,* Tetra International, 1975, Tetra Press, 201 Tabor Road, Morris Plains, NJ 07950.

Braemer, H. and Sheumann, I. *Tropical Fish,* Barron's, Munich, Germany, 1982.

Creatures in the Classroom, Multnomah County Intermediate Education District: Portland, Oregon, 1976. This comprehensive resource book can be purchased from: Curriculum Department, Multnomah County, Intermediate Education District, P.O. Box 16657, Portland, OR 97216.

Mills, D., *A Fishkeeper's Guide to Community Fishes,* Salamander Books, Ltd., 1984.

Mills, D., Vevers, G. *Tropical Aquarium Fishes,* Salamander Books, Ltd., 1982.

A Fish Story About Animal Attraction

New study about the quirky ritual of choosing a mate gives clue as to why some get away

New York

The male swordtail fish has a long sword-like appendage extending from its rear fin that seems to serve no purpose other than to attract females — but that attraction is not sex, researchers believe.

The big paddle-shaped tail may somehow resemble food, or it may merely attract the female's visual system, the researchers report. But in experiments that involved sewing fake tails on fish of a different species, they are shedding light on some curious quirks in evolution.

Researchers proved that the female preference for swords came before the evolution of the appendages by sewing plastic swords to the hindquarters of male platyfish, relatives of swordtail fish that have no swords.

Even though their partners naturally lacked swords, female platyfish overwhelmingly preferred the surgically enhanced males. The results of the study are being published today in the journal Science.

The study lends fuel to evolutionary biologists who maintain that females are attracted to flamboyant male traits not because they signify genetic advantages like strength or resistance to disease but because the features remind the females of something else.

The debate between the two schools of thought is among the liveliest in evolutionary biology.

"This is an outstanding study," said Mark Kirkpatrick, associate professor of zoology at the University of Texas in Austin. "It shows there are an awful lot of mating preferences that are established for reasons other than choosing the best possible mate."

Researchers want to know why females choose males with exaggerated traits

Biologists have long known that certain male traits, such as the peacock's gaudy tail, evolved over generations of selection by finicky females. But researchers have had difficulty finding what females are choosing when they opt for males with the most exaggerated versions of a particular trait.

In an effort to collapse vast stretches of evolutionary time into a single experiment, Alexandra L. Basolo, a researcher at the University of California at Santa Barbara, compared two closely related, brightly colored freshwater fish found in Mexico and Central America.

One species was the green swordtail; the male's body is about three inches long and has a three-inch sword-like extension from its rear fin. The other species was the southern platyfish; the males are an inch and a half long and lack tail swords.

The two species are members of the same genus, but the platyfish is considered more primitive, a kind of living progenitor of the green swordtail.

In past experiments, Basolo had shown that female swordtail fish preferred males with long swords over those with shorter ones. In the latest study, she tried to determine whether female platyfish might also be drawn to males with swords, even though their natural mates lack them.

To test the theory, she and a colleague who had surgical experience attached plastic swords to male platyfish. Each sword measured about an inch to match the platyfish's smaller body and was colored black and bright yellow, similar to the male swordtail's.

For the sake of comparison, other male platyfish were equipped with transparent swords that the females could not see.

The platyfish were anesthetized and the swords sutured on in a 40-second operation. A representative male of each group was then placed in different partitions of a tank, with a female swimming between them.

Every time a female platyfish was presented with a choice between a male with a colorful sword and a competitor with a transparent one, she invariably preferred the one with the colorful sword.

"Females would spend about 700 seconds with the sworded males, compared to only 400 seconds with the males that had the transparent swords," Basolo said.

To the researcher, the study encapsulates the origin and the evolution of a trait. She said she believed that among the green swordtails, a mutational event resulted in a male being endowed with a tiny sword. That male was then preferred by females because the sword appealed to an existing preference.

MORE ON THEMES

The word "themes" is used in many different ways in both ordinary usage and in educational circles. In the GEMS series, themes are seen as key recurring ideas that cut across all the scientific disciplines. Themes are bigger than facts, concepts, or theories. They link various theories from many disciplines. They have also been described as "the sap that runs through the curriculum," to convey the sense that they permeate through and arise from the curriculum. By listing the themes that run through a particular GEMS unit on the title page, we hope to assist you in seeing where the unit fits into the "big picture" of science, and how the unit connects to other GEMS units. The theme "Patterns of Change," for example, suggests that the unit or some important part of it exemplifies larger scientific ideas about why, how, and in what ways change takes place, whether it be a chemical reaction or a caterpillar becoming a butterfly. GEMS has selected 10 major themes:

Systems & Interactions	**Scale**
Models & Simulations	**Structure**
Stability	**Energy**
Patterns of Change	**Matter**
Evolution	**Diversity & Unity**

If you are interested in thinking more about themes and the thematic approach to teaching and constructing curriculum, you may wish to obtain a copy of our handbook, *To Build A House: GEMS and the Thematic Approach to Teaching Science.* For more information and an order brochure, write or call GEMS, Lawrence Hall of Science, University of California, Berkeley, CA 94720. (510) 642-7771. **Thanks for your interest in GEMS!**

Fish Facts

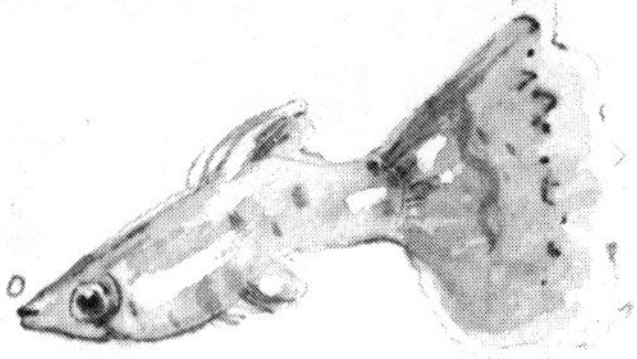

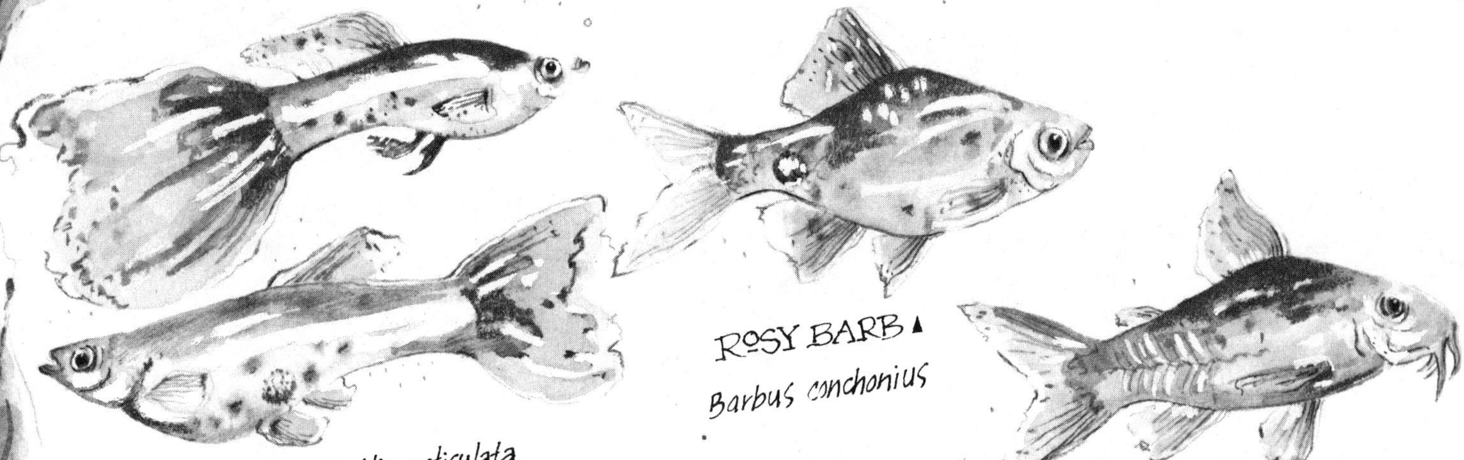

GUPPY ▲ *Poecilia reticulata*

ROSY BARB ▲ *Barbus conchonius*

BRONZE CATFISH ▲ *Corydoras aeneas*

Native to South America, guppies eat mosquito larvae and other insects at the surface of the water. Standard guppy flake food contains mosquito larvae, brine shrimp, other protein foods, and plants. Mated pairs of guppies breed frequently, and the female gives birth to live young. The male guppy's anal fin is rod-shaped and the female's is fan-shaped.

Native to northeastern India, rosy barbs travel together in schools. They can be fed on flake food or tablet food and will also nibble plants. The males have more red and black on their fins than the females, and turn a deep red color at breeding times. They breed frequently and will scatter their eggs around the aquarium.

Native to South America, bronze catfish travel in schools. They can be fed flake food or tablet food, but will also scavenge for food along the bottom and eat algae and plants. When breeding, the females will often place the fertilized eggs on plants.

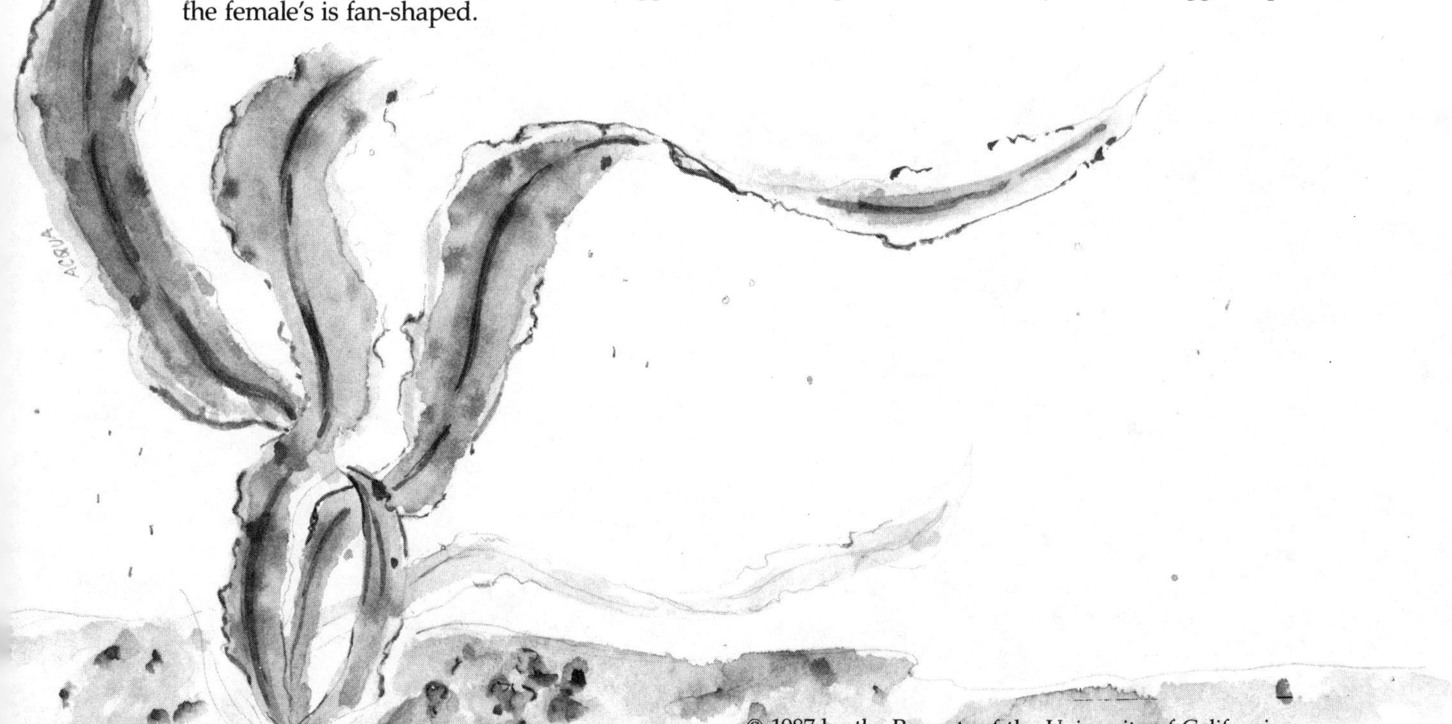

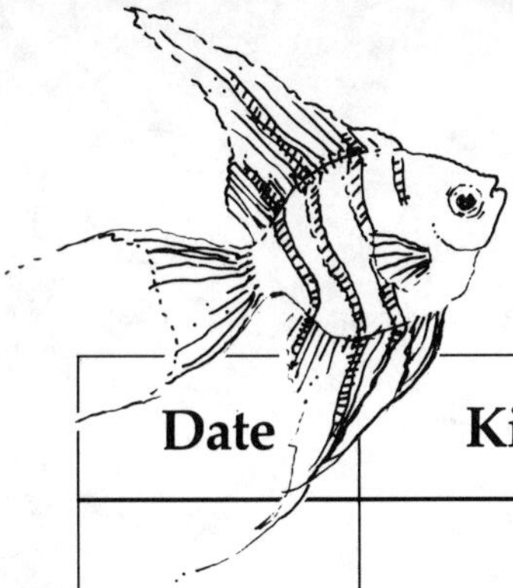

Name ______________________

Fish Observations

Date	Kind of Fish	Describe the Actions of the Fish

GEMS Guides

TEACHER'S GUIDES

Acid Rain
Grades 6–10

Animal Defenses
Preschool–K

Animals in Action
Grades 5–9

Bubble-ology
Grades 5–9

Buzzing a Hive
Preschool–3

Chemical Reactions
Grades 6–10

Color Analyzers
Grades 5–8

Convection: A Current Event
Grades 6–9

Crime Lab Chemistry
Grades 4–8

Discovering Density
Grades 6–10

Earth, Moon & Stars
Grades 5–9

Earthworms
Grades 6–10

Experimenting with Model Rockets
Grades 6–10

Fingerprinting
Grades 4–8

Global Warming
Grades 7–10

Height-O-Meters
Grades 6–10

Hide a Butterfly
Preschool–3

Hot Water & Warm Homes from Sunlight
Grades 4–8

Involving Dissolving
Grades 1–3

Liquid Explorations
Grades K–3

Mapping Animal Movements
Grades 5–9

Mapping Fish Habitats
Grades 6–10

More Than Magnifiers
Grades 6–9

Of Cabbages & Chemistry
Grades 4–8

Oobleck: What Do Scientists Do?
Grades 4–8

Paper Towel Testing
Grades 5–8

QUADICE
Grades 4–8

River Cutters
Grades 6–9

Vitamin C Testing
Grades 4–8

ASSEMBLY PRESENTER'S GUIDES

The "Magic" of Electricity
Grades 3–6

Solids, Liquids, and Gases
Grades 3–6

EXHIBIT GUIDES

Shapes, Loops & Images
all ages

The Wizard's Lab
all ages

SUPPLEMENTARY MATERIALS

GEMS Teacher's Handbook
GEMS Leader's Handbook
A Parent's Guide to GEMS
To Build a House: GEMS & the Thematic Approach to Teaching Science

---

● *Please contact the GEMS project for a descriptive brochure and ordering information.*

Write or call
GEMS
Lawrence Hall of Science
University of California at Berkeley 94720
(510) 642-7771